MÖSSBAUER ANALYSIS OF THE ATOMIC AND MAGNETIC STRUCTURE OF ALLOYS

MÖSSBAUER ANALYSIS OF THE ATOMIC AND MAGNETIC STRUCTURE OF ALLOYS

V.V. Ovchinnikov

Viva Books

New Delhi | Mumbai | Chennai | Kolkata | Bangalore | Hyderabad | Kochi

First Indian Edition 2008

VIVA BOOKS PRIVATE LIMITED

- 4737/23 Ansari Road, Daryaganj, New Delhi 110 002
  E-mail: vivadelhi@vivagroupindia.net, Tel. 42242200
- Plot No. 76, Service Industries, Shirvane, Sector 1, Nerul, Navi Mumbai 400 706
  E-mail: vivamumbai@vivagroupindia.net, Tel. 27721273, 27721274
- Jamals Fazal Chambers, 26 Greams Road, Chennai 600 006
  E-mail: vivachennai@vivagroupindia.net, Tel. 28290304, 28294241
- B-103, Jindal Towers, 21/1A/3 Darga Road, Kolkata 700 017
  E-mail: vivakolkata@vivagroupindia.net, Tel. 22836381, 22816713
- 7, Sovereign Park Apartments, 56-58, K. R. Road, Basavanagudi, Bangalore 560 004
  E-mail: vivabangalore@vivagroupindia.net, Tel. 26607409, 26607410
- 101-102 Mughal Marc Apartments, 3-4-637 to 641, Narayanguda, Hyderabad 500 029
  E-mail: vivahyderabad@vivagroupindia.net, Tel. 27564481, 27564482
- Beevi Towers, First Floor, SRM Road, Kaloor, Kochi 682 018, Kerala
  E-mail: vivakochi@vivagroupindia.net, Tel: 0484-2403055, 2403056

www.vivagroupindia.com

Published by arrangement with
Cambridge International Science Publishing
7, Meadow Walk, Great Abington
Cambridge CB1 6AZ, UK

ISBN-10: 81-309-0833-6
ISBN-13: 978-81-309-0833-5

Published by Vinod Vasishtha for Viva Books Private Limited, 4737/23, Ansari Road, New Delhi-110 002. Printed and bound by Anand Sons, Delhi-110 092.

Preface

I am pleased to welcome English-language readers and I would like to say several words about why I have written this book.

Undoubtedly, the Mössbauer effect in the period since its discovery has acquired a position as one of the most informative methods of examination of the atomic and magnetic structure of solids. However, of the large number of original publications, it is relatively difficult to extract fundamental principles representing the basis of these investigations. This must be carried out using several generalisations, and I have undertaken this task. I regarded it as necessary because it is clearly evident that the discovery of the Mössbauer effect has indicated a number of completely new possibilities in the examination of the atomic and magnetic structure of solids, in comparison with the possibilities provided by the conventional methods.

I participated in the development of non-diffraction Mössbauer methods of analysis of the atomic and magnetic structure of alloys [31,59,63,138,139,213 – FMM (Rus), 203,204, 330 – FTT, Pis'ma JTF (Rus), 68, 427 (Engl), and others], representing an alternative to the conventional methods based on the diffraction phenomenon (x-ray, electron and neutron). However, the above studies, published mainly in Russian journals, are usually not known to foreign readers.

The justification of the Mössbauer methods of the analysis of the atomic and magnetic structure of alloys required analysis of the problems associated with the listing of different non-equivalent positions of resonant nuclei (atoms) in these systems. In addition to this, it was necessary to develop methods of the analytical description of the probabilities of appropriate positions, in relation to the condition of the investigated objects (with rational restriction of the radius within which the differences in the interactions are taken into account). The probabilities of the positions specify the relative intensities of 'sub-spectra' which form the Mössbauer spectrum. Their number for alloys with a variable composition, including multiphase alloys, with imperfect long- and short-range great atomic and magnetic orders, and also of multilayered superlattices, is relatively large. Finally, it was necessary to describe the electrical and magnetic superfine interactions of resonant nuclei in the investigated positions (in the presence of atoms of different type in several closest coordination spheres of these nuclei) which determine the position of the individual lines in the composition of the sub-spectra.

The conclusions of the current theory of the interpretation of the results of indirect investigations are linked most directly with the problems of obtaining reliable information from the Mössbauer spectrum. This relates to the problem of well-founded construction of an entire class of permissible models and of selection, from this class, of the best model (in agreement with the error of the initial data). The final result are the optimum estimates of the physical parameters and of their errors. The most important role is played here by apriori information (fundamental physical laws, symmetric considerations, exact knowledge of apparatus functions, characterising the properties of apparatus, etc, so that it is possible to construct generalised models). The deviation from the principles of this theory results in inaccurate information obtained from the experiment on the structure of investigated objects and also estimates of the appropriate physical parameters and their errors. Part of the permissible models may be simply lost. Therefore, special attention is given to these problems in the monographs.

Being a nuclear-physical model, the Mössbauer effect provides information averaged out in respect of the ensemble of the individually absorbing resonant nuclei distributed in the crystal or amorphous solid.1 This results in many important cases in considerable advantages of the method in comparison with diffraction methods (x-ray, electron and neutron diffraction), in particular, in examination of the initial stages of atomic ordering, formation of disperse phases, nanocrystalline and amorphous materials. In these cases, the zones of coherent scattering are extremely small, and if they are smaller than 8–10 nm, diffraction effects simply do not form. For the same reason, because of the absence of measurable diffraction effects, the Mössbauer methods has advantages in the examination of the atomic and magnetic structure of materials with similar scattering properties of the components, low-symmetry superstructures and diluted solid solutions.

---

[1]At reduced temperatures $T < (0.4–0.6)T_{melt}$ the position of the large majority of atoms in solids during the lifetime of the excited state of the Mössbauer nucleus (10^{-11}–10^{-6} s) remains unchanged.

To help the readers not concerned professionally with the Mössbauer effect, and also students and investigators starting their work, I have provided in the book a brief introduction to the theory of the Mössbauer effect.

The metallic alloys which are the main subject in the monograph are very specific objects of investigation because of collectivisation of the electrons, the presence of variable composition phases, multiphase structure, etc. This is reflected in the complex nature of describing a large number of non-equivalent positions of Mössbauer nuclei (taking into account the variable atomic environment in several closest coordination spheres) and of the description of appropriate variations of their electrical and magnetic state. Consequently, these problems, starting with the general formulation to final equations, are examined in detail in chapters 2-4 of the book. Chapter 5 is concerned with the description and analysis of a large number of experimental investigations of the atomic and magnetic structure of alloys using the Mössbauer effect. Chapter 6 examines investigations of the Mossbauer spectroscopy of ion-doped metals and alloys (and also of silicon and some compounds). In these studies, attention was given to examining defect formation and other diverse and complicated processes induced by ion radiation.

The Appendices 1 and 2 are concerned with the methods of processing ('decoding') of the Mössbauer spectra. In Appendix 3 the accurate results, obtained by the author and confirming the excellence of the homogeneous (probability) and microdomain (topological) methods of description of the short-range order are described. The topological approach makes it possible to transfer from the short-range to long-range order with an increase of the dimensions of the ordered domains.

The latter is an important argument in discussing the justification of the application of the Mössbauer effect for examining not only the short-range but also long-range atomic order (regardless of the fact that each of the resonant nuclei, distributed in the crystal, 'senses' only its nearest environment).

I hope that the book will attract the attention of readers to these problems and will stimulate to a certain extent further development of the applications of the Mössbauer effect in the area of examination of the atomic and magnetic structure of the condensed media.

In the English edition, I have corrected a number of errors made in the original Russian version, and also made some additions and improvements in terminology.

In conclusion, I would like to express my gratitude to Cambridge International Science Publishing and, in particular, to Mr Riecansky, for the interest in my monograph and excellent work in publishing it.

V.V. Ovchinnikov

Preface to the Russian Edition

The method of nuclear gamma resonance, or the Mössbauer effect, discovered several decades ago, has been transformed into one of the most powerful methods of investigation of the structure of matter. Its experimental possibilities are continuously expanding. This relates to the selective indepth analysis of solids with registration of conversion and Auger electrons, optical, x-ray and gamma radiation. The applications, associated with the utilisation of synchrotron radiation, are being efficiently developed. Coherent effects are studied. The method is used widely for fundamental and applied investigations in chemistry, biology and solid state physics in order to examine the dynamics of atoms and the atomic and electronic structure of matter.

At the same time, analysis of the publications of the recent years indicates a crisis in the development of the method. This relates directly to the analysis of the atomic and magnetic structure of solids and is associated, in particular, with the absence in many cases important in practice of clear understanding of how to interpret unambiguously the results of Mössbauer experiments. Some fundamental problems relating to, for example, the relationship of the values of hyperfine magnetic fields on the atom nuclei with the values of the individual atomic magnetic moments have not as yet been explained. Theory bypasses many key problems relating to the formulation and interpretation of Mössbauer experiments. The experimenters often do not use efficiently theoretical results. So, unique information on the structure of condensed matter remains outside the examined areas. For example, almost all Mössbauer investigations of the long-range atomic order in the crystals are of qualitative nature, whereas the precision methods of analysis of superstructures using the Mössbauer effect were developed approximately 30 years ago. This is also combined with the crisis of classical methods of interpreting the results of investigations.

The Mössbauer effect provides no visual pattern of the structure of the crystal in the direct or reciprocal space, unlike a series of other methods. Nevertheless, in practice, the speculative interpretation of the obtained information is often used. This means that the individual special features of the Mössbauer spectra (for example, the well resolved maxima of resonant absorption) are intuitively attributed to special features of the structure and dynamic state of the investigated objects. The decomposition of the spectrum into arbitrary, at investigator's option, number of subspectra may be carried out using the method of computer fitting of the experimental data and the selection of the optimum values of the parameters of the subspectra.

It is clear that some important part of the information may be lost in this approach.

The aim of the monograph is to attract attention of investigators to a number of key problems whose solution is essential for further development of the Mössbauer methods of analysis of the atomic and magnetic structure of alloys, and also propose variants of the solutions of some of these problems on the basis of generalisation of the results of a large number of theoretical and experimental investigations, including the original studies by the author and his colleagues.

Chapter 1 is a brief introduction to the theory of the Mössbauer effect. This is dictated by the fact that the monographs, containing information from the theory of the Mössbauer effect, have already become a bibliographic rarity. Chapter 1 also gives an insight into the structure of Mössbauer spectra of a non-ideal crystal representing a superposition of the subspectra, corresponding to different non-equivalent states of the resonant nuclei in the solid, differing in the composition and symmetry of their nearest atomic environment.

At present, it is well known that in the experimental investigation of any object, in addition to the clear understanding of the theory of the method, it is necessary to utilise to the maximum extent the a priori information on the structure of the object, including all available data and theoretical representations. This is used as a basis for the construction of a class of permissible models, linking the structure of the investigated object with the results of observation (measurements). It is further determined whether the given class of the models contains at least one model, comparable with the existing experimental data (in respect of the accuracy of the description, matched with the accuracy of measurements), and this is followed by the selection of the most suitable model on the basis of the statistical criteria of agreement. It is this solely acceptable approach that makes it possible to control the entire procedure of analysis of the initial data from the beginning to the end and, in the final analysis, evaluate the reliability of the results, but unfortunately, it is practically not used.

The point is that the experimenter, having sufficient information on the theory of the method of nuclear gamma resonance, usually does not possess an adequate model reflecting the features of the investigated systems (in this case, we are concerned with a generalised model which takes into account the entire volume of a priori information, i.e., actually the class of the permissible models).

Chapter 2 is concerned with the problems of constructing such a model and interpreting Mössbauer spectra of alloys.

On the basis of the analysis of a priori information on the structure of a real crystal, it is shown that there may be constructed a class of permissible models of sufficiently universal nature, or a generalised model of the Mössbauer spectrum of the crystal, which provides for the possibility of its development for every particular case.

Such a model is presented for metallic alloys, taking into account the multiphase nature of these alloys, and a specific example is investigated. Attention is given to the problem of the quasi-continuous description of Mössbauer spectra which makes it possible to restore some relationships at the minimum a priori information.

Chapters 3 and *4* deal with the two most important aspects of the above-mentioned generalised model of the Mössbauer spectrum. They include the methods of description of the electrical and magnetic hyperfine interactions in metals and alloys (determining the position of the subspectra lines) and the methods of calculating the probabilities of non-equivalent configurations of the local environment of the resonant nuclei formed by the atoms of different type, determining the intensity of the subspectra. Indeed, in order to construct a generalised model of the Mössbauer spectrum of an alloy, in addition to data from the theory of the Mössbauer effect, only two items are required at the known or fixed dynamic parameters (in this monograph, no attention is given to the effects associated with the dynamics of the atoms): 1) to describe adequately the probabilities of various configurations of the local environment of resonant nuclei (in terms of the population of crystallographic positions in the presence of a long-range order, and of correlation moments in the presence of a short-range order), and 2) have a theory describing the electrical and magnetic hyperfine interactions of the nuclei of a resonant isotope surrounded by atoms of different type (the latter influence the spatial distribution of the charge and spin density of electrons).

Since the alloys have a number of specific properties (collectivisation of electrons, a wide spectrum of non-equivalent positions of resonant nuclei, the presence of defects of different type, multiphase structure, etc), the application of the Mössbauer effect for investigating metal alloys requires attacking a wide range of special problems.

In the area of experimental investigation and the theory of hyperfine interactions in metals and alloys, a number of outstanding studies have been conducted by: G.K.Wertheim et al. [26,35], M. Stearns, S. Wilson, A. Overhauser [32–34], R.E. Watson, A.J. Freeman [27,28], and others. The results are presented in paragraphs 3.1 and 3.2 of the book. In the studies by the author of the book, his teachers and colleagues I.N. Bogachev, V.S. Litvinov, S.D. Karakishev, F.A. Sidorenko [5,50,51,202–204], the data obtained in the above investigations were partially reproduced and supplemented, and new results have also been obtained.

Together with P.V. Gel'd [30, 31], the author of the present work for the first time analysed (to final equations and numerical values) the role of different mechanisms in the formation of hyperfine magnetic fields on the nuclei of the atoms of the matrix in ferromagnetic impurity crystals based on transition metals, and made generalisations, forming the basis of the second part of Chapter 3. Taking into account the main mechanisms of formation of the field of hyperfine interaction on the nuclei of the atoms of the matrix in diluted ferromagnetic alloys based on 3*d*-metals (RKKI-polarisation and repopulation of conduction electrons, Fermi contact interaction, lattice dipole contributions) equations have been obtained linking the strength H_{ef} of effective magnetic fields on the nuclei of the atoms of the transition metal-solvent, in different local environment formed by the atoms of the impurity, with the values of the *z*-projections of the individual atomic magnetic moments of the impurity and magnetic moments of the matrix atoms perturbed by the atoms of the impurity. The isotropic and anisotropic parts of the contributions to H_{ef} were discriminated. All this represents a basis for the description of the Mössbauer spectra of alloys (construction of a generalised model).

As a result, the method of determination of the magnetic moments of the atoms in alloys on the basis of the results of measurements of the concentration dependences of magnetisation and H_{ef} was justified. The data on Mössbauer and magnetic measurements were used to calculate the values of the *z*-projections of the magnetic moments of 2*sp*-, 3*sp*, 3*d*-, 4*d* and 5*d*-elements, dissolved in iron, indicating the strict periodicity with the change of the atomic number of the impurity. The best agreement with the results of neutron diffraction analysis was obtained for the elements of the 3*d*-period.

The equations, obtained for diluted solid solutions (it is assumed that the fraction of the impurity-impurity interactions does not exceed 1–2%) may be used as a starting point for describing hyperfine interactions in concentrated alloys. The approach used in [30, 31] may be regarded as eclectic, but it is the only relatively successful attempt to link the values of the node moments of greatly differing impurities with the effective magnetic field on the nuclei of the atoms of the matrix and the structure of the nuclear gamma-resonant spectra using a single equation. The result adds up small perturbations, determined by the effect of different mechanisms. Unfortunately, the author knows of nothing better which could be proposed in this direction.

Chapter 4 is concerned with the development of the second component of the generalised model of the Mössbauer spectrum: description of the relationship of the parameters of the spectra with the parameters of the long-range and short-range atomic order in the alloys. Different aspects of this problem have been examined more than 30 years ago in the studies of several independent research groups, mostly Russian scientists: R.N.Kuz'min, L.A. Losievskaya [86,64,66] (the long-range order and the short-range order on the background of the long-range order), P.L. Gruzin, Yu.L. Rodionov et al. [25,134], A.Brummer et al. [60] (short-range order), I.N. Bogachev, V.S. Litvinov, V.V. Ovchinnikov et al. [5, 59, 67–71] (long- and short-range order). In the studies [59, 67–71] the authors for the first time obtained the relationships taking into account the formation in alloys of ordered structures of the first, second and third rank. Also the effect of formation of the long- and short-range atomic order not only on the intensity but also on the width and position of the spectral lines was evaluated. The dependence of the mean value of the effective magnetic fields on the concentration and parameters of the short- and long-range atomic orders was obtained, taking into account the effect of an arbitrary number of coordination spheres.

In this Chapter, special attention is also given to the justification of the applicability of the Mössbauer effect having the local-nuclear nature (i.e., 'sensing' only the local atomic environment) for the examination of the short-range (including microdomain) and long-range atomic order in the alloys. The same problem is addressed in Appendix 3 in which the equivalence of the homogeneous ('probability') and microdomain short-range order models is proved for *AB* type alloys in which the atoms of one sublattice are surrounded by the atoms of another sublattice only. An exact relationship, linking the mean effective diameter of the antiphase domain with the parameters of the pair correlation has been determined.

The modelling of probabilities (by the Monte Carlo method) of the local atomic environment for the nuclei of a resonant isotope at the interphase boundaries of superlattices $\{A_m/B_n\}_k$ is also regarded.

Chapter 5 reviews the experimental investigations of the atomic and magnetic structure of alloys using the Mössbauer effect (mainly classical, and also those carried out in the last 10–15 years), directed to solving both fundamental and applied problems. The studies by the author of the book and his closest colleagues are used as an example for illustrating the principles of modelling the Mössbauer spectra, presented in previous chapters, and investigating the atomic (long-range and short-range) and magnetic order in the alloys. Special features of the investigations of non-collinear spin structures (following the studies by F.A. Sidorenko, B.V. Ryzhenko, A.V. Zaborov, B.Yu. Goloborodsky, et al.) and also magnetic superlattices (studies by V.V. Ustinov, L.N. Romashev, V.A. Tsurin, V.V. Ovchinnikov, et al.) have been considered.

Working at the Institute of Electrophysics of the Ural Branch of the Russian Academy of Sciences for the last 15 years, the author investigated the effect of the beams of accelerated ions on metals and alloys using the Mössbauer effect. In cooperation with Yu.E. Kreindel, G.A. Mesyats, high-speed non-thermal structural phase transformations self-propagating into the depth of the matter in the metastable media under the effect of high-power ion beams were revealed. These transformations are externally similar to the combustion and detonation phenomena, leading to the formation of unique electrical, magnetic, tribological and other properties of materials. These effects form a special group of the 'long-range effects' in ion irradiation.

On the whole, it is possible to construct a space-time hierarchy of the processes in implanting accelerated ions into matter. These are the processes in the zone of penetration of the ions and its immediate vicinity associated with ion doping, formation and migration of defects, and also long-range effects, based on the change of the structure and properties of materials far outside the range of the ions. The changes taking place in the structure and properties of metal targets may be investigated selectively over the depth using the currently available Mössbauer techniques.

Chapter 6 reviews Mössbauer investigations of ion-doped metals and alloys (including the studies by the author and his colleagues).

The results are presented of the investigations of the localisation and mobility of implanted impurities, the structure and stability of the resultant defects, the changes in the composition of the surface layer of the target, formation and dissolution of phases, etc. Attention is also given to the long-range effects which enable in practice to proceeding from ion beam modification of submicron to modification of submillimetre subsurface layers of the materials.

Appendices 1 and *2* are concerned with the formulation and methods of solving the *nonlinear parametric* (determination of the parameters of optimum models and their errors) and also *linear ill-posed* problems (restoration of the dependences from empirical data, without using physical models, with the minimum of a priori information). The last type of problems includes the inverse problems of spectroscopy. The solution of the inverse problem of Mössbauer spectroscopy makes it possible to transform the experimental spectrum to the distribution function of one or the other physical parameter (effective magnetic field, quadrupole splitting, energy of resonance).

The reason for the detailed examination in Appendix 2 of 'ill-posed' problems, unstable in the classic formulation, is that they are found more and more frequently in investigations associated with the analysis of the results of indirect observations (measurements). The information on the classical and non-classical methods of the restoration of relationships from empirical data, and also on the algorithms of solving unstable problems, utilising the regularisation concept, is absolutely essential for the accurate formulation and interpretation of Mössbauer experiments.

In the book, use is made of supplements to the monograph by V.S. Litvinov, S.D. Karakishev, V.V. Ovchinnikov [5], written by the author and, with the kind permission of V.S. Litvinov and S.D. Karakishev, the sections of Introduction and parts of Chapter 4 of the above monograph (some of them are written on the basis of the original studies in which the author took part). The author is very grateful for this. These sections are essential for maintaining the historical sequence and logics of presentation.

The author is grateful to G.A. Dorofeev, G.G. Amigud, N.A. Pervukhin, Sh. Daniyarov, V.A. Shabashov and V.A. Semionkin for joint work on the problems associated with the investigation of the atomic and magnetic structure of alloys which have been reflected in the monograph.

The author is also very grateful to his colleagues N.V. Gushchina, L.S. Chemerinskaya, and M.Yu. Yakhontova for their help in preparation of the manuscript, and also to the Russian Foundation for Basic Research for the financial support in publishing the book in Russia.

V.V. Ovchinnikov

CONTENTS

Chapter 1

BRIEF DESCRIPTION OF THE NATURE OF THE MÖSSBAUER EFFECT

MAIN PARAMETERS OF RESONANCE ABSORPTION SPECTRA

1.1. INTRODUCTION

The method of Mössbauer (nuclear gamma resonance) spectroscopy is based on the effect of resonant absorption of gamma quanta by the nuclei of atoms of crystals discovered in 1958 by R.L. Mössbauer [1, 2]. Let us discuss briefly the nature of this phenomenon, paying special attention to the relationship of the measured quantities with the structure and dynamic characteristics of solids.

The energy of the nuclei is quantised. In transition of a nucleus from the excited to the ground state, a γ-quantum with the energy $E = \omega\hbar$ ($\hbar = h/2\pi$) is emitted. Here $\hbar$ is the Planck constant. The most probable value of this energy for an infinite heavy free nucleus is equal to the difference of the energies of its ground and excited states: $E_0 = E_e - E_g$.[1] The inverse process corresponds to the absorption of the γ-quantum with the energy close to E_0.

In the excitation of a set of identical nuclei on the same level, the energy of the emitted quanta is characterised by a scatter

[1]In this case, there are no losses of energy for recoil. For free nuclei there are also no effects of displacement and splitting of nuclear levels characteristic of condensed media, as a consequence of the interaction of nuclei with intramolecular and intracrystalline electrical and magnetic fields.

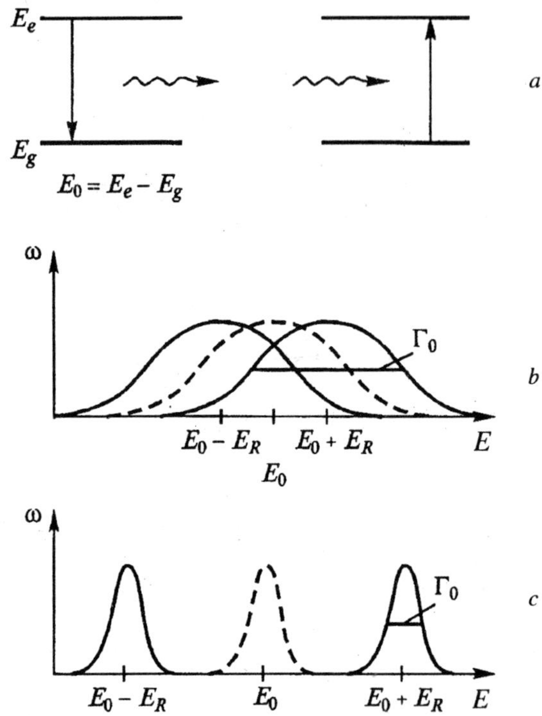

Fig. 1.1. Diagram illustrating quantum transitions with radiation and absorption of electromagnetic quanta (*a*) and the form of radiation and emission lines in optical (*a*) and nuclear (*b*) cases.

around the mean value E_0. The form of the emission line, i.e. the density of probability of radiation of a γ-quantum with energy E, is determined by the Breit–Wigner equation (Fig. 1.1):

$$\omega(E) = \frac{1}{\pi} \frac{(\Gamma_0/2)}{(E - E_0)^2 + (\Gamma_0/2)^2}, \tag{1.1}$$

where Γ_0 is the width of the emission line at half of its height (natural width of the line), equal to the indeterminacy of the energy of the excited state of the nucleus. In accordance with the relation of the uncertainties $\Gamma_0 = \hbar/\tau$ or $\Gamma_0 = \hbar \ln 2/\tau_{1/2}$, where τ and $\tau_{1/2}$ are the mean lifetime and the half-life of the excited state of the nucleus. The mean lifetime of the excited state for different nuclei is usually in the range $10^{-11} \div 10^{-6}$ s, which corresponds to the width of the line of $\sim 10^{-5} \div 10^{-10}$ eV.

The contour of the absorption line is described by the same relationship as the contour of the emission line (1.1). Of course, the effect of resonance absorption is observed in experiments only when the lines of emission and absorption greatly overlap.

Investigations were carried out into the phenomenon of the resonance absorption of electromagnetic radiation of the optical range in which the optical quanta, emitted in transition of the electrons of excited atoms to the underlying electronic levels, are absorbed in the resonant manner by the substance containing atoms of the same type. The phenomenon of optical resonance absorption can be clearly seen in, for example, sodium vapours.

Unfortunately, the phenomenon of resonance nuclear absorption is not detected in the case of free nuclei. This is caused by the fact that the model of heavy nuclei (atoms) when the energy losses in recoil in relation to Γ_0 are not large, holds for optical resonance and cannot be used at all for nuclear resonance. The gamma quanta, emitted in nuclear transitions, are characterised by a considerably higher energy, tens and hundreds of keV (in comparison with several electron volts for the quanta of the visible range). For the comparable values of the lifetime and, correspondingly, similar values of the natural width of the electronic and nuclear levels in the nuclear case the recoil energy plays a considerably more significant role in emission and absorption: $E_R = p^2/2m = E^2/2mc^2$, where $p = E/c$ is the recoil pulse of the nucleus equal in the modulus to the pulse of the emitted γ-quantum, m is the mass of the nucleus (atom). It may easily be calculated that in the optical case the order of magnitude of E_R is the same as that of Γ_0. Consequently, optical resonance takes place on free (non-bonded) atoms. However, in the nuclear case $E_R \gg \Gamma_0$ there is no resonance on the free nuclei (see Fig. 1.1 *b* and *c*).

In 1958, Rudolf Mössbauer, examining the absorption of γ-quanta, emitted by the isotope ^{191}Ir, observed in the ^{191}Ir crystal, in contrast to the predictions of classic theory, an increase of the extent of scattering of γ-quanta at low temperatures ($T \approx 77$ K) [1, 2]. He showed that this effect is associated with the resonant absorption of γ-quanta by the nuclei of the atoms of ^{191}Ir and published an exhaustive explanation of the nature of this phenomenon.

Classic studies on the theory of the Mössbauer effect and its most important applications are found in subject collections of studies [3, 4].

1.2. THE NATURE AND PROBABILITY OF THE MÖSSBAUER EFFECT

The nature of the detected phenomenon may be explained qualitatively on the basis of the Einstein model,[1] according to which a crystal, containing N atoms, is represented by the set of $3N$ harmonic oscillators with the same frequency $\omega = \omega_E$. The state of the solid is given by the quantum numbers n_i ($i = 1, 2, ..., 3N$) and the respective energies of the oscillators $E_i = \hbar\omega(n_i + 1/2)$ ($n_i = 0, 1, 2, ...$). When the atom is fixed in the crystal and the recoil energy is lower than the energy required for knocking out the atom from the lattice site,[2] the energy and recoil pulse are divided between the excited phonons (lattice oscillations) and the crystal, as a single quantum system. The energy which the γ-quantum exchanges with the phonons may assume discrete values: $0, \pm h\omega, \pm 2h\omega, ...$

Each of these processes is characterised by the probability: f_0, f_1^+, f_1^-, f_2^+, f_2^-... Consequently, there is a fully determined probability $f = f_0$ of processes taking place without any change of the oscillation energy of the lattice. The recoil energy of the crystal as an integral unit is negligible ($E_R = p^2/2M$, p is the pulse of the γ-quantum, M is the mass of the entire crystal). Thus, with the given probability, one can detect the phonon-free and recoil-free emission or absorption of γ-quanta by the atom nuclei.

The Mössbauer effect is the phenomenon of nuclear resonance absorption of γ-quanta in cases when the losses of energy in recoil and excitation of the phonons (in both emission and absorption) are not present because of the given reasons.

The probabilities $f = f_0$ and $f' = f_0'$ are the probabilities of phonon-free and recoil-free emission and absorption of γ-quanta (by the source and the absorber) or, according to the currently used terminology, also the probabilities of the Mössbauer effect. In a general case, f and f' differ. The probability of the Mössbauer effect is especially high at $T = 0$ K. For an absorbing nucleus ^{57}Fe in metallic iron in the vicinity of Kelvin zero $f' \approx 0.92$.

In the experiments with the Mössbauer effect, measurements are not taken of the emission lines (or absorption lines) but of the curves of resonance absorption (i.e. the Mössbauer spectra, see equations (1.3)–(1.5)). The unique applications of the method of nuclear gamma resonance in solid state chemistry and physics of

[1] For elements of a more accurate theory, based on the Debye model, see below.
[2] I.e. the emission (or absorption) of a γ-quantum is not accompanied by radiation damage and the atom remains in its position in the lattice.

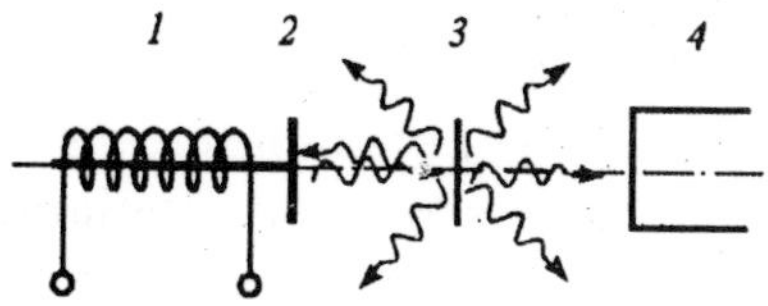

Fig. 1.2. Diagram of the Mössbauer experiment: 1) electrodynamic vibrator setting different values of the speed v of the source; 2) Mössbauer source (for example, ^{57}Co); 3) an absorber containing the spectra of the Mössbauer isotope (^{57}Fe); 4) detector of gamma quanta passed through the absorber (usually a proportional counter or a photoelectronic multiplier).

solids are determined by the fact that the width of the individual resonance lines, forming the Mössbauer spectrum, is smaller than the energies of the magnetic and electrical interactions of the nucleus with the surrounding electrons. The Mössbauer effect is an efficient method of examining a wide range of phenomena affecting these interactions.

The simplest scheme for the examination of the Mössbauer effect in transmission geometry includes a source, an absorber (a thin specimen of the investigated material), and a detector of gamma rays (Fig. 1.2; see also Fig. 6.1 in chapter 6).

The source of gamma rays should have specific properties: a long half-life of the mother nucleus (as a result of the breakdown of this nucleus, the nucleus of the resonance isotope in the excited state is formed), the energy of Mössbauer transition should be relatively low (to ensure that the recoil energy does not exceed the energy required for the displacement of the atom from the lattice site), the emission line should be narrow (this results in high resolution), and the probability of phonon-free emission should be high. For Mössbauer spectroscopy of iron alloys these requirements are satisfied by ^{57}Co with a half-life of 270 days which is a mother isotope for the ^{57}Fe isotope. Usually, sources with an activity of $1 \div 100$ mCi are used.

The source of γ-quanta is produced in most cases by introducing a Mössbauer isotope into a metallic matrix by diffusion annealing. The material of the matrix should have a cubic lattice (in order to prevent quadrupole splitting of the line), and should be dia- or paramagnetic (magnetic splitting of the nuclear levels is prevented). The effects of hyperfine splitting of the lines are examined in section 1.3.

Absorbers are thin ($0 < C_a \leq 6$) specimens (see equations (1.4), (1.5)) in the form of foils or powders. When determining the required thickness of the specimen, it is necessary to take into account not only the content of the Mössbauer isotope in the material but also the probability of the Mössbauer effect. In the case of pure iron, the optimum thickness is ~20 μm, i.e. approximately 0.16 mg/cm^2 of isotope ^{57}Fe. The optimum thickness is the result of a compromise between the necessity to operate with a thin absorber and obtain a strong effect of absorption.

For the recording of the γ-quanta, transmitted through the specimen, scintillation and proportional counters are most often used.

Formation of the spectrum of resonance absorption (or the Mössbauer spectrum) requires variations of the resonance conditions. For this purpose, it is necessary to modulate the energy of the γ-quanta. The currently used modulation method is based on the Doppler effect (in most cases, the source of γ-quanta moves in relation to the absorber).

The energy of the γ-quantum as a result of the Doppler effect changes by the value:

$$\Delta E = \frac{v}{c} E_0 \cos\theta, \tag{1.2}$$

where v is the absolute speed of movement of the source in relation to the absorber; c is the velocity of light in vacuum; θ is the angle between the direction of movement of the source and the direction of emission of γ-quanta.

Since angle θ in the experiments has only two values, $\theta = 0$ and $\pi/2$, then $\Delta E = \pm v/c$ (the positive sign corresponds to approach, the negative sign to the movement of the source away from the absorber).

In the absence of resonance, for example, when the absorber does not contain the nucleus of the resonant isotope or when the Doppler velocity is very high ($v \to \pm\infty$, which corresponds to disruption of resonance because of the very large change of the energy of the γ-quanta), the maximum part of radiation, emitted in the direction of the absorber, falls into the detector positioned behind the absorber. The signal from the detector is amplified and pulses from the individual γ-quanta are recorded by the analyser. Usually, recording is made of the number of γ-quanta in the same period of time at different values of v. In the case of resonance, the γ-quanta are absorbed and re-emitted by the absorber in arbitrary directions (Fig. 1.2). The fraction of radiation, falling on the detector, decreases in this case.

In the Mössbauer experiment, the dependence is analysed of the intensity of radiation passed through the absorber[1] (the number of pulses recorded by the detector) on the relative velocity of the source v. The effect of absorption is determined by the relationship:

$$\varepsilon(v)=[I(\infty)-I(v)]/I(\infty), \tag{1.3}$$

where $I(v)$ is the number of γ-quanta, recorded by the detector in the specific period of time at the value of Doppler velocity v (the experiments are carried out using a discrete set of velocities v_i); $I(\infty)$ is the same at $v \rightarrow \pm\infty$, when there is no resonance absorption. The relationships $I(v)$ and $\varepsilon(v)$ specify the shape of the curve of resonance absorption (Mössbauer spectrum) in absolute and relative units. The values of velocity v, required for the measurement of the curves of resonance absorption of alloys and compounds of iron, are in the range ± 10 mm/s.

The value of the resonance effect may be presented in the following form [5]:

$$\varepsilon(y)=\chi f\left\{1-\frac{1}{\pi}\int_{-\infty}^{\infty}\frac{1}{1+x^2}\exp\left[-\frac{\sigma_0 nf'}{1+(x+y)^2}\right]dx\right\}, \tag{1.4}$$

where $y = 2E_0 v/\Gamma_0 c$ (v is Doppler velocity, c is the velocity of light in vacuum); χ is the fraction of the resonance γ-quanta in radiation of the source; $x = 2(E-E_0)/\Gamma_0$; f, f' are the probabilities of emission and absorption of the γ-quanta without recoil; σ_0 is the absorption cross-section at accurate resonance (the nuclear constant for the given Mössbauer isotope); n is the number of the atoms of the isotope per 1 cm^2 of the absorber.

Quantity $C_a = \sigma_0 nf'$, which is present in the numerator of the exponent (1.4) and is independent of energy, determines the effective thickness of the absorber for the resonance quanta. If there is no self-absorption in the source, then for $0 < C_a \leq 6$ (this absorber is referred to as thin), the Mössbauer spectrum may be approximated by the Lorentz curve:

[1]Here and in the rest of the book, if not specified otherwise, measurements in transmission geometry are referred to.

$$\varepsilon(v) = \Gamma^2 \varepsilon(0) / (\Gamma^2 + 4v^2), \tag{1.5}$$

where $\varepsilon(0) = \chi f C_a/2$, $\Gamma = 2\Gamma_0$. Expression (1.5) may be obtained from equation (1.4) if we use in the integrand the first two terms of the expansion of the exponent into a series in respect of C_a powers.

The probability of the Mössbauer in effect is determined by the phonon spectrum of the crystal. In Debye approximation, this probability is given by the equation [6]:

$$f' = \exp[-2W(T)], \tag{1.6}$$

where $W(T)$ is the Debye–Waller factor:

$$W(T) = \frac{3E_R}{k_B\Theta}\left[\frac{1}{4} + \left(\frac{T}{\Theta}\right)^2 \int\limits_0^{\Theta/T} \frac{\varphi}{e^{\varphi} - 1} d\varphi\right]. \tag{1.7}$$

At low temperatures ($T \ll \Theta$), probability f' reaches values close to unity (at the absolute zero $f' = \exp[-3E_R / (2k_B\Theta)]$, where k_B is the Boltzmann constant), and at high temperatures ($T \gg \Theta$) it is very low. Equation (1.7) shows that, with other conditions being equal, the probability of phonon-free absorption and emission is higher in crystals with a high Debye temperature. The latter determines the strength of the interatomic bond.

The classic theory of the Mössbauer effect makes it possible to obtain a simple and easy-to-interpret-representation of the Debye–Waller factor [7]:

$$2W(T) = 4\pi^2 \langle X^2 \rangle / \lambda^2, \tag{1.8}$$

where $\langle X^2 \rangle$ is the mean square of the amplitude of oscillations of the nucleus in the direction of emission of the γ-quantum, λ is its wavelength.

Equations (1.7) and (1.8) show that the probability of the effect is determined by the spectrum of elastic oscillations of the atoms in the crystal lattice. The intensity of the Mössbauer line is high if the amplitude of oscillations of the atoms is not high in comparison with the wavelength of γ-quanta, i.e. at low temperatures. In this case, the absorption and emission spectrum consists of a narrow resonance line (phonon-free processes) and a wide component, determined by the variation of the vibrational states of the lattice in emission and absorption of γ-quanta (the width of the latter is six orders of magnitude higher than the width of the resonance line).

The anisotropy of the atomic bond in the lattice determines the anisotropy of the amplitude of oscillations of the atoms and, consequently, the different probability of phonon-free absorption in different crystallographic directions. In the case of single crystals it is therefore possible to measure not only average but also angular dependences of f and f' and obtain estimates of the appropriate force constants.

In the approximation of the thin absorber, the probability of phonon-free transitions is proportional to the area below the curve of resonance absorption which may be calculated from the equation[1]:

$$S = (\chi/2)\pi\sigma_0\, n\, f\, f'. \tag{1.9}$$

Nuclear gamma resonance may be used for determining the vibrational properties of the solid-state lattice or impurity atoms in this lattice. A more suitable experimental parameter in this case is the area of the spectrum S because it is the integral characteristic and does not depend on the shape of the spectrum of emission of resonance quanta and self-absorption in the source [6]. This area is retained in splitting of the spectrum into several components as a result of hyperfine interactions. The probability of the Mössbauer effect may also be determined from the measurements of the temperature (relativistic) shift of the Mössbauer spectrum, caused by the Doppler effect of the second order (for this subject, see [5]).

1.3. EFFECTS OF DISPLACEMENT AND SPLITTING OF THE LINES. PARAMETERS OF MÖSSBAUER SPECTRA

As indicated by the relationships (1.3) and (1.5), the simplest spectrum of resonance absorption of the thin absorber is repre-

[1]For an absorber of arbitrary thickness, the area of the experimental spectrum can be determined from the equation [5, p.9]:

$S = \chi f \pi C_0 e^{C_0/2}\left[I_0\left(i(C_0/2)\right) + I_1(i(C_0/2))\right]$, where $I_0(i(C_0/2))$ and $I_1(i(C_0/2))$ are Bessel functions of the imaginary argument of the zeroth and first order.

sented by a single line of the Lorentz shape (Lorentz function is substracted from the constant):

$$I(v) = I(\infty)\left[1 - \frac{a}{1 + 4/\Gamma^2 (v - C)^2}\right], \tag{1.10}$$

where $a \equiv \varepsilon\ (C) = \chi\ n\ \sigma_0\ ff'/2$ is the strength of the effect at maximum absorption; constant C in the denominator takes into account the fact that the absorption and emission lines are usually displaced in relation to each other (even in the absence of recoil and excitation of the phonons). The shift is caused by a number of effects, examined in this paragraph. The intensity of the radiation, passed through the absorber, is minimum at maximum absorption. The curve (1.10) resembles the section of an inverted bell with the

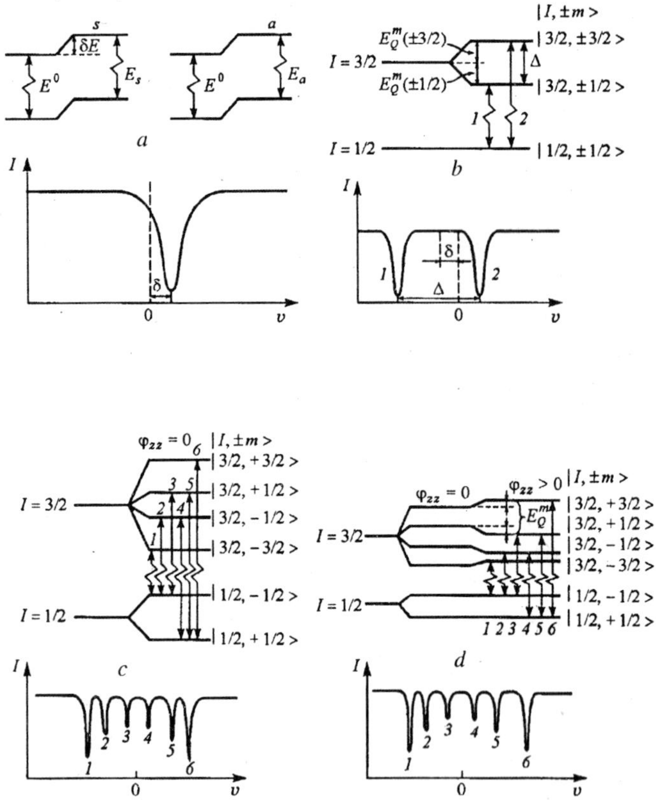

Fig. 1.3. Displacement and splitting of energy levels of ^{57}Fe nucleus and the Mössbauer transmission spectra in isomeric (chemical) shift (a), quadrupole interaction (b), magnetic dipole interaction (c) and combined interaction (magnetic dipole + relatively weak quadrupole interaction) (d) [8].

minimum at $v = C$ (Fig. 1.3*a*) [8].

The shape of the experimental line may differ from the Lorentz shape and its width does not correspond to the ideal case of the thin absorber because of apparatus distortions and a large number of physical effects (see paragraph 1.4).

The shift of the centre of gravity of the Mössbauer spectrum may be caused by the following reasons [9]:

1. The difference in the energy of zero oscillations of the lattices of the source and the absorber;
2. Temperature red shift;
3. Different isomeric shift (for the source and the absorber);
4. Differences in hyperfine interactions (for groups of atoms).

The shifts of the first two types are of a common nature: they are associated with the relativistic reduction of the mass of the nucleus in radiation, resulting in an increase of the general energy of the lattice and, consequently, a decrease of the energy of the γ-quantum (red shift).

Attention should also be given to the fact that the shift of the centre of gravity of the spectrum forms at the variation of the volume of the alloy as a result of the appropriate re-normalisation of the wave functions. For the alloys and compounds of iron $\delta = -1.4\ \Delta V/V$ mm/s [10].

Because of the interaction of resonant nuclei with electrical or magnetic intracrystalline fields, the $(2I+1)$-fold degeneration of the energy levels (I is the spin of the nucleus) may be completely or partially suppressed. The spectrum is transformed in this case into the superposition of several lines. The so-called hyperfine structure (HFS) of the Mössbauer spectrum forms.

We examine in greater detail the nature of the main effects, resulting in the shift and splitting of the nuclear levels.

Temperature shift δ_T of the Mössbauer spectrum

The variation of the energy of the γ-quanta is observed at the complete structural identity of the source and the absorber, if their temperature differs (absolute or Debye). This results in the temperature shift of the Mössbauer spectrum (the Doppler shift of the second order). In the Debye approximation, the temperature shift (mm/s) may be calculated from the equation:

$$\delta_T = -2.715\times10^{-4}\left[\Theta + 8T(T/\Theta)^3 \int_0^{\Theta/T} \frac{\varphi}{e^{\varphi}-1}d\varphi\right]. \qquad (1.11)$$

This shift is a relativistic effect determined by the difference in the mass of the nucleus in the ground and excited states and by the respective variation of the kinetic energy of the nucleus [11].

The temperature shift of the lines is relatively small. For example, in the case of pure iron at $\Delta T = 150$ K, the order of δ_T corresponds to the natural width Γ_0. For α-Fe, the temperature shift of the spectrum may be calculated from equation (1.11) using the value of the effective Debye parameter $\Theta_D = 470$ K [12, page 122].

If the chemical interaction of the atoms may be regarded as temperature-independent, from the measurements of the Doppler shift of the second order we obtain information on the dynamics of the atoms (it should be mentioned that the mean quadratic velocity of the vibrational movement of the nucleus in the crystal is the function of the force interaction of the atoms).

The temperature shift does not change the form of the absorption spectrum (1.10) and only provides a contribution to the shift of the spectrum in relation to zero velocity (or the reference absorber).

Isomeric[1] (chemical) shift of the Mössbauer spectrum

The nature of this shift is determined by the electrostatic interaction of the nucleus with the surrounding electronic charge (electrical monopole interaction).

The isomeric shift of the spectral line of absorption in relation to the irradiation line is determined by the equation [7, 12]:

$$\delta = (2\pi/3) Z e^2 \left(\langle r_e^2 \rangle - \langle r_g^2 \rangle \right) \left(|\Psi_a(0)|^2 - |\Psi_s(0)|^2 \right) \tag{1.12}$$

and represents a product of two co-multipliers. The first of them contains only nuclear parameters: the charge of the nucleus Ze and the mean quadratic radii of the nucleus $\langle r_g^2 \rangle$, $\langle r_e^2 \rangle$ in the ground and excited states, and the second includes the atomic parameters: the density $\rho_{a,s}(0) = e|\Psi_{a,s}(0)|^2$ of the electronic charge at the point of position of the nucleus ($r = 0$) in the absorber (a) and in the source (s). For the isotope ^{57}Fe, the mean quadratic radius of the nucleus in the ground state is higher than in the excited state. Therefore, isomeric shift δ decreases with increase of the charge density of the charge on the nucleus in the absorber, and increases with density decrease.

[1]The term 'isomeric' is used due to the fact that the shift depends on the difference of the radii of the nucleus in the ground and isomeric (excited) states.

The shift provides information on the distribution of the external *s*- and *d*-electrons of the resonant atom. The electron density on the nucleus is generated only by the electrons with the *s*-symmetry which have spherical wave functions; *d*-electrons (like *p*- or *f*-) have a 'knot' (*i.e.* zero density) on the nucleus, but influence the isomeric shift indirectly, screening the *s*-electrons. With an increase of the number of *d*-electrons the isomeric shift for ^{57}Fe increases, and vice versa.

For the alloys and compounds of iron, the effect of the *p*-electrons is considerably weaker than the effect of *d*-electrons and in many cases is not taken into account.

Equation (1.12) specifies the isomeric shift of the absorber in relation to the resonance isotope in the matrix of the source. When comparing the isomeric shifts on the same isotope in different alloys and chemical compounds, the magnitude of these shifts should be expressed in relation to some single etalon. The shift in relation to the etalon can be easily determined if the shifts for the reference and the investigated substance in relation to some source are available.

Since the internal *s*-shells do not undergo significant changes during the formation of solid solutions and chemical compounds, the isomeric shift of ^{57}Fe provides information on the redistribution of external *s*- and *d*-electrons. Consequently, it will also be referred to as the *chemical* shift. The variation of the chemical shift in a series of compounds or alloys provides direct information on the nature of variation of the chemical bond.

The isomeric shift results in the shift of the 'centre of gravity' of the Mössbauer spectrum, specified by the relationship (1.10), in relation to the etalon absorber.

It should be mentioned that the interpretation of the isomeric shifts for the alloys of transition metals encounters considerable difficulties because of the absence of a strict electronic theory for them.

Quadrupole splitting E_Q

The nuclei with the spin 0 and 1/2 are characterised by the spherically symmetric form. Their electric quadrupole moment is consequently equal to zero. The nuclei with the spin $I > 1/2$ do not have spherical symmetry and are characterised by the quadrupole moment Q which differs from zero. The interaction of the quadrupole moment with the gradient of the electrical field (generated by the surrounding electrical charges) results in partial removal of degeneration in respect of the magnetic quantum number and splitting of the levels of the nucleus into sub-levels (the dependence of energy on the square of the magnetic quantum number m_I appears). The value of the splitting of the excited level of the nucleus ^{57}Fe with the spin $I = 3/2$, situated in the field with the axially symmetric gradient of the electrical field eq', is determined by the expression [7, 12] (Fig. 1.3*b*):

$$\Delta E_Q = 1/2e^2 q'Q(1-\gamma_\infty), \tag{1.13}$$

where $(1-\gamma_\infty)$ is the antiscreening factor, which makes it possible to take into account the effect of partially filled shells of the Mössbauer atom on the gradient of the electrical field. This gradient, determined by the charges of the ions surrounding the Mössbauer atom, is not equal to zero if the symmetry of the environment of the resonant atom is lower than cubic. In the approximation of point charges, the components of the tensor of the gradient may be determined from the equation [7]

$$V_{ik} = \frac{1}{e}\sum_\alpha \frac{e_\alpha}{R_a^3}\left(3\mathbf{n}_i^\alpha \mathbf{n}_k^\alpha - \delta_{ik}\right) \tag{1.14}$$

where $\mathbf{n}$ is the unit vector in the direction R_a; δ_{ik} is the Kronecker symbol.

In the case of considerable splitting of the level ($\Delta E_Q \gg \Gamma$), the absorption spectrum is a well-resolved doublet (see Fig. 1.3*b*). Since the gradient of the electrical field decreases with increase of the distance in proportion to $1/R^3$, the strongest effect on the value of the gradient is exerted by the atoms of the nearest neighbourhood. For example, in solid solutions, extensive quadrupole splitting of the resonance level forms when an interstitial impurity appears in the nearest neighbourhood of the Mössbauer atom.

The theoretical form of the doublet may be represented by the superposition of two Lorentz lines:

$$I(v) = I(\infty)\left[1 - \frac{a/2}{1+4/\Gamma^2\left(v-C-\Delta E_Q/2\right)^2} + \frac{a/2}{1+4/\Gamma^2\left(v-C+\Delta E_Q/2\right)^2}\right]. \tag{1.15}$$

As in the case of isomeric shift, E_Q is determined by the product of the nuclear and atomic co-multipliers. The value Q of the resonant nucleus is a nuclear constant and is determined from independent experiments. The values of Q for the nuclei of different isotopes are presented in, for example, a monograph by Shpinel' [12]. The atomic multiplier $q'(1-\gamma_\infty)$ may be calculated on the basis of theoretical considerations. Comparing its value with the distance between the lines of the experimental doublet, we can identify the position of the atoms in the lattice of a solid solution. Comparison of the experimental results with the calculated results, for example, in the model of point charges, may be used to determine the charge state of the impurity in the crystal.

Magnetic dipole splitting

If the atomic nucleus in the energetic state E, characterised by the magnetic moment $\boldsymbol{\mu}$, differing from zero, is placed in a magnetic field $\boldsymbol{H}$, which is constant with time, the energy of the nuclear state changes by the value [13]:

$$\Delta E = -(\boldsymbol{\mu H}) = -(m_I / I)\mu H, \tag{1.16}$$

where I is the spin of the nucleus in the state with energy E, m_I is the magnetic quantum number, assuming $2I+1$ values: $I, I-1, \ldots, -I$. Since in contrast to the case of the electrical quadrupole interaction, the variation of energy ΔE is proportional to the first-degree m_I, degeneration in respect of the magnetic quantum number is completely removed. In the absence of the magnetic field in the experiments on nuclear gamma resonance, measurements are taken of the transition between the states E_g, I_g and E_e, I_e, and in the presence of the field between E_g, I_g, m_{Ig} and E_e, I_e, m_{Ie}. The sampling rules for the magnetic quantum number: $\Delta m_I = 0, \pm 1$, result in the case of nucleus ^{57}Fe ($E_g = 0$, $I_g = 1/2$, $m_{Ig} = \pm 1/2$ and $E_e = 14.4$ keV, $I_e = 3/2$, $m_{Ie} = \pm 1/2, \pm 3/2$) in six resolved transitions and in the formation, in the Mössbauer spectrum of magnetically ordered substances, of six individual absorption lines (Zeeman nuclear effect) (Fig. 1.3*c*).

Using the value μ_g (^{57}Fe) = 0.0903 ± 0.0007 μ_n[1] obtained in [14] by the NMR method, and the value μ_e (^{57}Fe) = 0.153 ± 0.004 μ_n, measured using the Mössbauer effect, Hanna et al [15] determined the strength of the field on the Fe nucleus in pure iron at room temperature: H(^{57}Fe) = 333 ±10 kOe.

The intensities of the lines of the Zeeman sextet of the

[1] μ_n is nuclear magneton.

magnetically ordered substances, containing the Mössbauer isotope ^{57}Fe, are related in the case of a thin absorber as 3:z:1:1:z:3, where $0 \leq z \leq 4$. Parameter z characterises the relative intensity of the transitions $-1/2 \rightarrow -1/2$, $+1/2 \rightarrow +1/2$ (for the second and fifth line of the sextet) and is the function of the angle between the direction of the beam of γ-quanta and the axis of the magnetic field. For polycrystaline specimens, on the condition of the equal probability of different directions of magnetisation in magnetic domains and the isotropic nature of the Debye–Waller factor, the mean value $\langle z \rangle = 2$[6].

The magnetic field on the ^{57}Fe nucleus in pure iron is anti-parallel in relation to the magnetic moment of the atom. This is associated with the fact that the main contribution to the effective field comes from the exchange polarisation of s-electrons of the internal shells of the atom by the resultant spin of $3d$-electrons [13]. The exchange interaction of s- and d-electrons determines the attraction of the electrons in parallel orientation of the spins and repulsion in the case of anti-parallel orientation. This results in the formation of the spin density of the s-electrons on the nucleus of the atom different from zero. The magnitude of this contribution is given by the expression [7]:

$$H = +\frac{8\pi}{3}\mu_B \sum_n \left[\left|\Psi_{n\uparrow}(0)\right|^2 - \left|\Psi_{n\downarrow}(0)\right|^2 \right], \tag{1.17}$$

where $|\Psi_{n\uparrow}(0)|$ and $|\Psi_{n\downarrow}(0)|$ are the densities of the s-electrons of the n-shell with the spin parallel and antiparallel to the magnetic moment of the atom.

In pure ferromagnetic iron, in addition to polarisation of the shells of the internal and external s-electrons, there are also other sources of the magnetic field on the nucleus. The contribution to the magnetic field is provided by the orbital moment of the electrons. According to the data in [7, 12], in metallic iron, the strength of the magnetic field, generated by the non-frozen orbital moment of $3d$-electrons, is equal to ~+70 kOe. Another source of the field is the contribution from the magnetic moments of the adjacent atoms, regarded as magnetic dipoles. In the case of the cubic crystals, consisting of identical atoms, this contribution is equal to zero. In iron and its alloys, the strength of the effective magnetic field is determined by the degree of polarisation of conductivity electrons by nodal (localised on the sites of the lattice) magnetic moments and, in addition to this, by the degree of hybridisation of conductivity electrons with 3*d*-electrons of the iron atom (see paragraph 3.3)

As indicated by equation (1.16), the magnitude of magnetic splitting of the nuclear levels and, correspondingly, the distance between the lines of the sextet are determined by the product of the constant nuclear μ and variable atomic H co-factors. Consequently, it is possible to measure the fields on the nuclei of the atoms of magnetic materials, examine their formation mechanism, and also investigate the effect on the effective magnetic field on the nucleus of factors such as composition, temperature, pressure, superposition of the external fields, etc.

The theoretical form of the Mössbauer spectrum in the presence of magnetic splitting of the nuclear levels may be represented by the superposition of the Lorentz lines:

$$I(v) = I(\infty)\left[1 - \sum_{i=1}^{6} \frac{a_i}{1 + 4/\Gamma_i^2 (v - C - A_i H)^2}\right], \qquad (1.18)$$

where A_i are the coefficients taking into account the fine structure of the energetic levels of the nucleus H, given for ^{57}Fe by the vector (+0.5; +0.289; +0.079; –0.079; –0.289; –0.5).

It should be mentioned that the magnetically-ordered alloys are often characterised by combined (magnetic and electrical) hyperfine interaction. There is no general solution of this problem (the calculation of the position of the energy levels of the nucleus). For the case of the axially symmetric tensor of the gradient of the electrical field, the hyperfine structure of the nuclear levels and the form of the Mössbauer spectrum in the combined interaction are shown in Fig. 1.3*d*. In many cases important for practice, for example, for substitutional solid solutions with cubic symmetry, the shift of the lines of the spectrum because of quadrupole splitting (as a result of local distortion of cubic symmetry, caused by the differences in the effective charge and the radius of the components) is very small and can be ignored to the first approximation.

1.4. THE WIDTH OF THE RESONANCE LINE

The width of the resonance line in an ideal case of a thin absorber (see (1.5)) is equal to the double natural width $2\Gamma_0$ of the excited level of the nucleus ($\Gamma_0 = \hbar/\tau$, where τ is the mean lifetime of the nucleus in the excited state). The actual experiments are characterised by the apparatus broadening of the line, determined by the characteristics of the given specific Mössbauer spectrometer (the level of vibrations, linearity, stability, etc), the broadening, determined by self-absorption in the source and the absorber because of their finite thickness [6] and, finally, the broadening associated with physical reasons linked with the investigated subject. The latter include the dynamic effects, determined by the movement of atoms [13], and also the effects associated with the presence of a wide spectrum of the states of the Mössbauer nuclei in the crystal because of the variation of the local atomic and electronic environment (in this case, the broadened resonant curve is the superposition of the closely spaced lines displaced in relation to each other or partially split lines). The broadening of the Mössbauer line may be caused by the high density of the point defects and the dislocations [16].

1.5. THE STRUCTURE OF RESONANCE ABSORPTION CURVES OF NON-IDEAL CRYSTALS

The relationships (1.10), (1.15) and (1.18) describe the form of the Mössbauer spectra of the crystals where all atoms are characterised by the same crystallographic, chemical and, correspondingly, electrical and magnetic states. However, even in strictly ordered crystals and chemical compounds, the atoms of the same element may be situated in different structural positions, have different valency and structure of the external electronic shells.

The number of different states of the atoms rapidly increases for non-stoichiometric and even stoichiometric but partially disordered phases. The non-ordered solid solutions, to which alloys of metals are often related, are characterised by a large number of different states of atoms. The resonant atoms in these systems may be found in a large variety of non-equivalent positions.

The real alloys may be characterised by a wide spectrum of different states of resonant atoms (and their nuclei) because of the difference of the configuration of the nearest neighbourhood of these atoms by chemical elements and defects in the radius of several closest coordination spheres. The presence of a high density of defects, dislocation nuclei, interphase and grain boundaries may result in changes of the state of a large number of atoms.

The states of the Mössbauer nuclei in different phases of the alloy also differ.

Because of the difference in the parameters of hyperfine interaction for the resonant atoms situated in different non-equivalent positions in the lattice, the Mössbauer spectra of real crystals often represents a very complicated superposition of responses from a large number of individually absorbing resonant nuclei.

Therefore, the experimental spectra contain very important (often unique) information on the topography and dynamics of the atoms of the crystal. This information can be obtained only if we know the relationship between the parameters of the Mössbauer spectrum and the parameters of the solid, i.e. when the form of the curve of resonance absorption is available as a function of the parameters relating to the atomic and electronic structure and dynamics of the atoms of the crystal.

When analysing the relationships presented in section 1.2, it may be concluded that the parameters, relating to the dynamics of the atoms and the parameters defining the number of the resonant nuclei in different non-equivalent states in the crystal, determine the amplitude of respective Lorentzians. However, the number of non-equivalent states determines the number of subspectra in the composition of the Mössbauer spectrum, and the parameters of hyperfine interactions determine the position of the lines of individual subspectra on the energy scale.

As an example, Fig. 1.4 shows the Mössbauer spectra of pure iron (*a*) and of an iron alloy with 8.25 at.% of Mn (*b*) quenched from 820°C, with the random distribution of iron and manganese atoms in the lattice sites. The graph presents the spectrum of the alloy as superposition of three Zeeman sextets of the lines, similar to the spectrum of pure iron but differing in the values of the effective magnetic field.

As indicated by the classic studies by Stearns, Wertheim and their colleagues (a detailed examination of these studies is presented in chapter 3), the sextet of the lines with the maximum value of the field $H_0 \approx H_{Fe}$ corresponds to the iron atoms[1] whose first coordination sphere does not contain the impurity atoms (manganese in the present case). The sextets H_1 and H_2 correspond to the atoms of ^{57}Fe which have respectively 1 and 2 manganese

[1]We gather information from the ^{57}Fe nuclei only, however the sample containing resonant isotope atoms completely reflects the properties of the sample of all iron atoms.

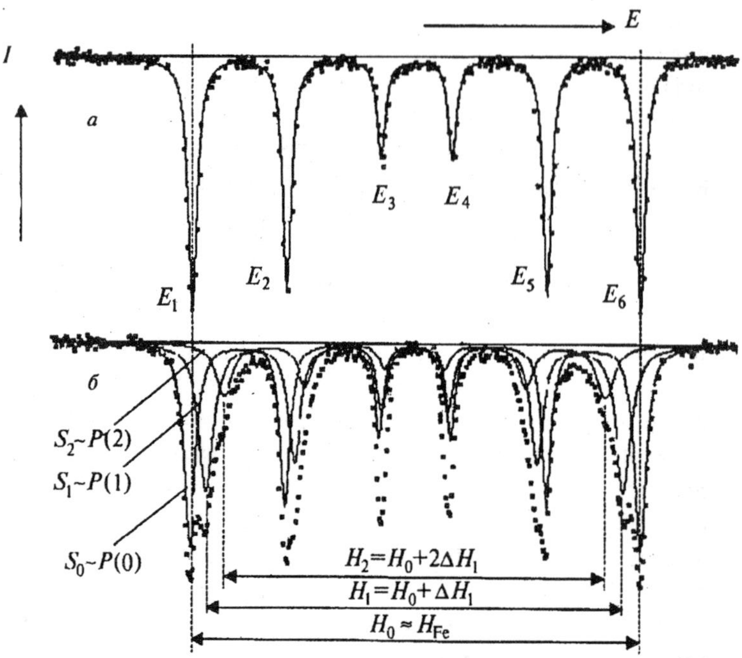

Fig. 1.4. Mössbauer spectra of pure iron (*a*) and Fe+8.25 at.% Mn (*b*) (T_{meas} = 300°C); S_0, S_1, S_2 are the areas defined by the most intensive sextets of lines in the absorption spectrum; ($P(0)$, $P(1)$, $P(2)$ are the probabilities of iron atoms being surrounded by l = 0, 1, 2 nearest Mn atoms); ΔH_1 is the contribution to H_{ef} from the Mn atom in the first coordination sphere of ^{57}Fe ($\Delta H_1 < 0$).

atoms among the nearest neighbours.

The rule of additivity of the contributions is fulfilled for diluted alloys, namely: $H_1 = H_0 + \Delta H_1$ and $H_2 = H_0 + 2\Delta H_1$ (in a general case, $H_l = H_0 + l\Delta H_1$, where ΔH_1 is the contribution to the effective magnetic field from the Mn atom in the first coordination sphere). The intensity of the three discussed sextets of the lines is proportional to the probabilities $P(l)$ of the Fe atom being surrounded by different numbers l = 0, 1, 2,..., z of the atoms of the impurity (z is the coordination number for the first coordination sphere in the BCC lattice). The identical effect of the atoms of the impurity is also detected for the isomeric shift: $\delta_0 \approx \delta_{Fe}$, $\delta_1 \approx \delta_0 + \Delta\delta_1$, $\delta_2 \approx \delta_0 + 2\Delta\delta_1$, and so on, where $\Delta\delta_1$ is the contribution to the isomeric shift from the atoms of the impurity of the first coordination sphere. The values $\Delta\delta_i$ (where i is the number of the coordination sphere), expressed in the values of Doppler velocity, are very small. Correspondingly, the visually detected shift of the components of the spectrum (the sextets of the lines in the present case) is also small. The mathematical processing of the spectra makes it possible to determine the values of ΔH_i and $\Delta\delta_i$ and evaluate their errors.

In the non-ordered alloy, the probability $P(l)$ is determined by the binomial distribution:

$$P(l) = C_z^l c^l (1-c)^{z-1}, \qquad (1.19)$$

where c is the concentration of the substitutional impurity (for the previously mentioned alloy c = 0.0825), C_z^l is the number of combinations from z in respect of l.

A more accurate model considers in the same manner the weaker effect of the second, third, etc. coordination spheres whose effect does not provide visually resolved lines in the Mössbauer spectrum, and is manifested externally as the broadening (and additional shift) of the lines.

Analysis of the hyperfine interactions and also the principles of modelling and decoding the resonance absorption spectra are discussed in chapters 2, 3 and 4, and in Appendix 1 in this book. They describe the problems relating to the examination of the atomic and magnetic structure of the alloys. Special attention is given to the methods of determination of the individual atomic magnetic moments of the atoms and also the parameters of the long- and short-range atomic orders in metallic alloys.

Chapter 2

INTERPRETATION OF MÖSSBAUER SPECTRA OF ALLOYS

The quantitative Mössbauer analysis of metals and alloys leads unavoidably to the problem of 'decoding' of γ-resonant spectra. The complexity of this problem may differ in relation to the nature and state of investigated objects. The strict formulation and solution of this problem requires the application of information from the theory of the Mössbauer effect, the physics of the real crystal, applied mathematics, and also knowledge of special algorithms and elements of programming. These problems have been discussed only briefly in the literature and require special examination

2.1. PRELIMINARY COMMENTS

The formation of the pattern of resonant absorption of γ-quanta by the crystals, is exerted contributed to by a large number of absorbing atoms which in a general case are not chemically, crystallographically, etc. equivalent. The structure of the Mössbauer spectra is consequently directly linked with the real structure of the solid. Analysis of the results of nuclear gamma resonance measurements in all stages require the application of a priori representations regarding the structure of the real crystal, which is one of the guarantees of the reliable interpretation of the experimental data.

As already mentioned, the Mössbauer effect is observed in one of the isotopes of a chemical element. Since different isotopes are chemically equivalent with a high degree of accuracy, all the parameters of interaction of the individual isotopes, included in the composition of the element and, consequently, their equilibrium macro- and microdistribution in the crystals are regarded as identical, and the extracted information is regarded as specific for the element as a whole.

It should be mentioned that in the experiments with the Mössbauer effect, as in any physical measurement, we obtain information averaged out over a specific period of time. In the Mössbauer measurements it is the lifetime of the excited state of the nucleus (from 10^{-11} to 10^{-6} s for different isotopes). The result of averaging of the effect of magnetic interactions on the position of the energy levels of the nucleus is also determined by the period of nuclear Larmor precession which is important in the measurement of hyperfine magnetic fields on the atom nuclei. Here, we do not give attention to the problems associated with diffusion jumps and collective displacements (such displacements may be detected in the vicinity of phase transition points [17,18]) of atoms during the periods of time not exceeding the lifetime of the excited state of the nucleus. Consequently, it is possible to introduce the concept of the *non-equivalent positions* of the resonant nuclei in the crystal which remain constant during the lifetime of the nuclear level. It should be mentioned that the evaluation of the frequency of diffusion jumps in the processes of self-diffusion and diffusion of substitutional impurities in different matrices at temperatures up to $0.7 \div 0.8\ T_m$ indicates the constancy of the position of the majority of the atoms during the lifetime of the excited state of the nucleus. In order to understand the nature of the formation of the structure of Mössbauer spectra, characteristic of alloys and compounds, it is necessary to examine in greater detail the problem of non-equivalent positions of the resonant nuclei in the crystal.

2.2. NON-EQUIVALENT POSITIONS OF RESONANT NUCLEI

The non-equivalent positions of the resonant nuclei in a real crystal are linked with special features of the structure of the crystal: the type of symmetry of the crystal lattice, defects, interfaces, difference in the composition of the nearest environment of the Mössbauer nuclei with the atoms of the chemical elements. Thus, in the physical sense, a special feature is the position in the vicinity of the point defects, dislocation nuclei, the grain boundaries, interphase boundaries, on the surface of the solid. Evidently, the positions of the atoms of the Mössbauer isotope in different phases of the alloy are also non-equivalent.

The problem of separation of the non-equivalent positions of the resonant nuclei is directly linked with the problem of the extent of interatomic interaction in solids, i.e. with the problem of the determination of the maximum distance at which the difference in the composition and geometry of electronic and ion environment is tangible.

Investigations of magnetic diffusion scattering of neutrons, nuclear gamma resonance and nuclear magnetic resonance indicates that the effects of perturbation of the charge $\Delta\rho(R)$ and spin $p\uparrow(R) - p\downarrow(R)$ density of the conductivity electrons, caused by the charge and localised magnetic moment of the impurity, decrease relatively rapidly, asymptotically, at $R \to \infty$, correspondingly as $\cos(2k_F R + \varphi)/R^3$ and $\cos(2k_F R)/R^3$, where R is the distance to the impurity, k_F is the pulse of the electron on the Fermi surface. However, it should be remembered that these perturbations are of the oscillating nature and, consequently, the effects of the atoms of the impurity of the first, second and even third coordination spheres may be found comparable. The quantum-mechanic calculations of the energy of atomic interaction $U(R)$, directly characterising the extent of the interaction, lead to the identical equation: $U(R) \sim A\cos(2k_F R + \varphi)/R^3$ [19]. The magnitude of the perturbation of the localised magnetic moments of the matrix in the vicinity of the atoms of the impurity also evidently decreases in proportion to $1/R^3$ [20].

In the studies published in [21,22], concerned with the Mössbauer examination of the oscillations of spin density in iron alloys, it was possible to resolve the effect of the first six coordination spheres and, consequently, it is possible to take into account very small changes in the state of the resonant atoms in different positions. In the solution of specific problems, we can confine ourselves to considering the interaction only with the nearest and next neighbours of the resonant nuclei. Obviously, in this case it is desirable to have at least an integral estimate of the effect of remote coordination spheres. In a number of cases, this effect may be taken into account indirectly.

Prior to examining the problem of the role of different non-equivalent groups of the atoms in the formation of the structure of γ-resonance spectra, it is important to note another significant circumstance, associated with the special features of Mössbauer investigations: we can define two cases differing in the nature of distribution of the atoms of the resonant isotope in the crystal: 1) 'natural' distribution (a suitable example are alloys melted by the conventional method; the corresponding element in this case may be enriched with the Mössbauer isotope); 2) the selective introduction of the resonant isotope – diffusion on crystal lattice defects (dislocations, grain boundaries), introduction of the isotope on the surface, etc. Possibly, the combination of these two cases, when the resonant isotope is additionally introduced, for example, on defects of the alloy naturally containing the Mössbauer nuclei. These comments relate to a certain extent to alloys whose components contain Mössbauer nuclei, and also to impurity nuclei (nuclei–probes).

2.3. THE SUPERPOSITION PRINCIPLE

Non-equivalent groups of resonant nuclei

In the natural distribution of the atoms of the resonant isotope in the alloy, the fraction of the Mössbauer nuclei, situated in the vicinity of the defects of the crystal lattice (point defects, nuclei of the dislocations, grain boundaries, etc) with the usual density of the latter, is small, as indicated by respective estimates. Consequently, the decisive importance is attributed to by the differences in the composition and geometry of the environment of the resonant nuclei by the atoms of the chemical elements, included in the composition of the alloy.

The results of examination of the local perturbation of the charge and spin density of the electrons in diluted [21, 22] and concentrated [23] solutions, including the use of the nuclei–probes [24], indicate the relatively wide range of application of the model of the pair interactions. In particular, the latter indicates that the difference in the state of the resonant nuclei is determined mainly only by the number of the atoms of different type in each of the several nearest coordination spheres and, to a considerably smaller degree, by the variation of their mutual distribution in these spheres.[1]

The application of the model of pair interactions makes it possible to find the consecutive and relatively general solution of the problem of decoding the experimental Mössbauer spectra which, evidently, may be used as the starting point for different improvements. Let us analyze these approximations for a single-

[1]In subsequent chapters of the book, we hypothesized the validity of the given assumption which is less rigid than the assumptions made in the model of the pair interaction and the model of central interaction.

phase alloy. We consider an alloy containing M components (the atoms of the resonant isotope are included in the composition of one of them). It is assumed that the quantities l_{ij} determine the number of atoms of the i type in the j-th coordination sphere of the selected resonant nucleus. In order to describe in full the state of the selected Mössbauer atom in the approximation of the paired interactions taking into account N nearest coordination spheres, it is necessary and sufficient to specify the matrix:

$$\begin{pmatrix} l_{11} & l_{12} & \cdots & l_{1N} \\ l_{21} & l_{22} & \cdots & l_{2N} \\ \cdots & \cdots & \cdots & \cdots \\ l_{M1} & l_{M2} & \cdots & l_{MN} \end{pmatrix}, \tag{2.1}$$

reduced to: $\|l_{ij}\|_{M,N}$. In this case, $\sum_{i=1}^{M} l_{ij} = z_j$ (z_j is the coordination number of the j-th sphere), i.e. any of the numbers in the column may be expressed by other numbers. The total number of the independent values l_{ij} is consequently equal to $(M - 1)N$. This comment enables us to omit the last line in the matrix (2.1), assuming that M is the type of element containing resonant atoms, i.e. it is possible to characterise the state of the resonant nuclei by the number of other elements in several nearest coordination spheres.

It is obvious that in the model of N spheres, the number of non-equivalent positions of the resonant nuclei L, equal to the number of different matrices, $\|l_{ij}\|_{M,N}$, or, which is the same, $\|l_{ij}\|_{M-1,N}$, is finite. Namely:

$$L = \prod_{j=1}^{N} C_{z_j+M-1}^{M-1}, \tag{2.2}$$

and since in the actual cases $L \ll n$ (n is the number of resonant nuclei in the specimen), the same matrix of type (2.1) corresponds to a large number of resonant atoms. In other words, there are L non-equivalent groups of the atoms of the resonant isotope.

Every non-equivalent group of the Mössbauer atoms is

characterised therefore by the respective matrix $\left\| l_{ij} \right\|_{M-1,N}$. In the case of a binary alloy, it is the line: l_1, l_2, l_j,..., l_N (the first index is omitted), where l_j is the number of atoms of the second component in the *j*-th coordination sphere of the resonant nucleus. Thus, at $N = 6$, one of the groups is formed by, for example, the atoms, characterised by the matrix (0, 2, 1, 1, 3, 2). No atoms of the impurity are found in the first coordination sphere, two atoms are in the second sphere, and so on. In this case, the total number of the non-equivalent groups L, according to equation (2.2), is equal to $(z_1+1)\ (z_2+1)\ldots\ (z_6+1)$. For example, in the case of the BCC lattice taking into account the effect of the six spheres $L = 9 \times 13 \times 7 \times 25 \times 9 \times 13$.

Different non-equivalent groups are characterised by the number of resonant nuclei which belong to them. Since the number of the Mössbauer nuclei in the groups depends on the volume of the specimen, it is more convenient to use the relative (instead of absolute) numbers of the groups: in the limit – the probabilities of the non-equivalent positions of the resonant nuclei in the crystal. If different matrices $\left\| l_{ij} \right\|_{M-1,N}$ are numerated in an arbitrary order, for example, using the index (k = 1, 2,..., L), then

$$P_k = n_k \left(\sum_{l=1}^{L} n_l \right)^{-1}, \quad \sum_{l=1}^{L} n_l = n, \tag{2.3}$$

where P_k is the probability of k-th non-equivalent position; n_k and n are, respectively, the number of the Mössbauer atoms in the k-th non-equivalent group, and the total number of these atoms. In this case it is evident that $\sum_{k=l}^{L} P_k = 1$.[1]

Superpositional structure of the spectrum

Discussion of the problem of participation of the different non-equivalent groups in the formation of the resultant structure of the Mössbauer spectrum should be started from the following comment. The non-equivalent positions of the resonant atoms in the crystals differ in the parameters of chemical, electrical and magnetic

[1]It should be remembered that, for example, for a binary alloy $\sum_{k=1}^{L} P_k = \sum_{l_1=0}^{z_1} \sum_{l_2=0}^{z_2} \cdots \sum_{l_N=0}^{z_N} P(l_1, l_2, \ldots, l_N) = 1$. A shortened version will be used in examining general problems.

interaction which, because of the interaction between the electronic and nuclear subsystems, is reflected in the variation of the energies of nuclear transitions and, although the order of magnitude of the latter is $10^{-9}\div10^{-6}$ eV, the extremely high sensitivity of the Mössbauer effect makes it possible to register these differences. The presence of the non-equivalent groups of the atoms thus determines the presence of completely or partially resolved components in the Mössbauer spectrum.[1]

As indicated by equation (1.9), the integral intensities of the components of the spectrum in the approximation of the thin absorber should be determined by the relationships $s_k = n_k f'[(\chi/2)\sigma_0 f]$, i.e. taking equation (2.3) into account:

$$s_k \sim P_k f_k'. \qquad (2.4)$$

In particular, the latter indicates that there is a mutually unambiguous correspondence between the total set of non-equivalent positions of the resonant nuclei in the crystal and the multitude of partial 'sub-spectra', forming the resultant γ-resonance spectrum.

The values of f_k in the model of the central interaction may be expressed by means of the force constants λ and λ', characterising the strength of the bond between the atoms of the matrix and the impurity. The results of [25], in which the relationship of f_k' with λ and λ' was investigated, indicate that in the majority of cases we should not expect any large differences in the individual values of f_k'. In most cases, $f_k' \approx$ const. Experimental estimates of the degree of accuracy of this assumption are not known to the author.

On the basis of the above examination it is possible to formulate the following assumptions: 1. Each non-equivalent group of the resonant atoms is related to the partial component[2] in the Mössbauer spectrum and vice versa; 2. The Mössbauer spectrum of the thin absorber in the approximation f_k' = const represents a sum of the components whose integral intensities are proportional to the probabilities of the positions, i.e. $P_1/s_1 = P_2/s_2 = \ldots = P_L/s_L$. The latter claim is the principle of linear superposition often used in practice.

Taking these assumptions into account, it is easy to evaluate the number of components forming the Mössbauer spectrum. In a

[1] In some cases, there is no high-resolution and the variations of energies may be obtained only as a result of the special computer processing of Mössbauer spectra.

[2] The individual component ('sub-spectrum') may represent both a single line and also a set of lines in splitting of nuclear levels

general case, this number is equal to L (see equation (2.2)). However, normally a preliminary estimate of P_k is made for different k ($k = 1, 2, \ldots, L$). The integral intensities s_k, corresponding to low values of p_k, may be rejected if the condition $\sum_j P_j << 1$ is fulfilled (summation in respect of rejected j). Usually, it is assumed that the given sum should not exceed ~0.01÷0.03. In fact, this means that the Mössbauer spectrum may be represented by the superposition of L' components. In this case, L' is usually considerably smaller than L.

It should be mentioned that in connection with the above considerations, it is quite evident that it is necessary to decode the Mössbauer spectrum in accordance with the a priori considerations regarding the structure of the real crystal. So, the arbitrary selection of the number of spectrum components, for example, on the basis of external features: the number of well-resolved peaks in the experimental spectrum, etc, without preliminary estimate of P_k may obviously lead to wrong results.

The above examination may be easily generalised for the case of a multi-phase alloy. The ratio of the areas of the components s_{k_l} in this case is evidently equal to the ratio of the components: $\gamma_l P_{k_l} f_l$,[1] where γ_l is the fraction of the atoms of the resonant isotope in l-th phase of the alloy ($k_l = 1, 2, \ldots, L_l$). Here

$$\sum_{k_l=1}^{L_l} P_{k_l} = 1 \quad \text{and} \quad \sum_{l=1}^{G} \gamma_l = 1 \quad (G \text{ is the number of phases}).$$

Since the values of s_{k_l} are directly linked with the values of γ_l and P_{k_l}, in each specific case we can obtain the preliminary estimate of the principal possibility of examination of different processes and phenomena using the Mössbauer effect. For example, if the spectrum is represented by the superposition of 'sub-spectra' corresponding to the individual phases of the alloy, then it is evident that any phase may be detected only on the condition that the fraction of the Mössbauer atoms in this phase γ_l is not too small. A rational boundary may be represented by the value $\gamma_l \approx 0.01$, i.e. approximately 1%.[2]

[1]In a general case, f'_{k_l} should be considered.

[1]It should be mentioned that the formation of the new phase, not containing the nuclei of the Mössbauer isotope, accompanied by the variation of the composition of the initial phase, may be recorded indirectly on the basis of the variation of the shape of the γ-resonance spectrum.

2.4. GENERALISED FORM OF THE RESONANCE ABSORPTION CURVE

The above examination enables us to transfer to the problem of the analytical description of the form of the curves of resonance absorption of single-phase and multiphase alloys. Let us consider two possible methods of describing the Mössbauer spectrum: discrete and continuous (or quasi-continuous).

Discrete description

The concept of this description is directly linked with the existence of non-equivalent groups of resonant nuclei in the crystal. Thus, in the case of a single-phase alloy, the form of the γ-resonance spectrum may be defined as follows:

$$I(v) = x_1\left[1-\sum_{j=1}^{L'} P(j, x_2, x_3, \ldots, x_s)\, x_{s+1}^{(j)} F\left(v; x_{s+2}^{(j)}, x_{s+3}^{(j)}, \ldots, x_p^{(j)}\right)\right], \tag{2.5}$$

where $I(v)$ is the intensity, corresponding to Doppler velocity v; x_1 is the count at infinity; $P(j;\, x_2, x_3, \ldots, x_s)$ is the probability of the j-th position being the function of the parameters $x_2, x_3, \ldots, x_s$, linked with the special features of distribution of the atoms in the alloy; for a weak dependence of f on $j\, x_{s+1}^{(j)} = \frac{\pi}{2}\chi n \sigma_0 f f'$ (see (1.9)); $x_{s+2}^{(j)}, x_{s+3}^{(j)}, \ldots, x_p^{(j)}$ are the parameters characterising the shift and splitting of the lines: isomeric shifts, the effective fields on the nuclei of the atoms, etc.[2] The function $F\left(v; x_{s+2}^{(j)}, x_{s+3}^{(j)}, \ldots, x_p^{(j)}\right)$ for the given value of j determines the individual partial 'sub-spectrum': single line or several lines of hyperfine splitting whose form is given in a general case by equation (1.4). Here, $F\left(v; x_{s+2}^{(j)}, x_{s+3}^{(j)}, \ldots, x_p^{(j)}\right)$ for different j should be normalised in such a manner that in the limiting case of a thin absorber:

$$\int_{-\infty}^{\infty} F\left(v; x_{x+1}^{(j)}, x_{x+2}^{(j)}, \ldots, x_p^{(j)}\right) dv = 1 \quad (j = 1, 2, \ldots, L'). \tag{2.6}$$

[1]Or the fundamental characteristics by means of which these parameters are expressed.

It should be mentioned that the values of the parameters x_1, $x_2,...,x_s$, $x_{s+1}^{(j)},x_{s+2}^{(j)},...,x_p^{(j)}$ are not available prior to the experiment. The methods of determination of these parameters on the basis of Mössbauer measurements are described in Appendix 1.

The given form of writing $I(v)$ reflects the specific features of alloys as objects of investigation, characterised by a large number of non-equivalent positions of the resonant nuclei (in contrast to, for example, pure metals and stoichiometric compounds). In fact, the Mössbauer spectrum of alloys often represents a complicated superposition of a large number of elementary components.

Consider a specific example. In the absence of quadrupole and magnetic dipole splitting of the nuclear levels, the form of the Mössbauer spectrum of the non-ordered single-phase alloy on the condition that the absorber is sufficiently thin, may be given by the expression:

$$I(v_k)=x_1\left[1-\sum_{j=1}^{L'}\frac{2}{\pi}\frac{P_j x_{2j}/x_{3j}}{1+4/x_{3j}^2\left(v_k-x_{4_j}\right)^2}\right], \tag{2.7}$$

where $x_1 = I(\infty)$, $x_{2j}=\frac{\pi}{2}\chi n\sigma_0 ff'$, x_{3j} and x_{4j} are respectively the values of the width and isomeric shift of the individual components. This is illustrated by Fig. 5.2 (in chapter 5).

The functions, giving the form of the individual lines in the approximation of the thin absorber in accordance with equation (1.5) are of the Lorentz type and satisfy the condition (2.6). The distribution of the probabilities in the case of complete disordering of an alloy is described by the Bernoulli scheme. In this case, P_j are not functions of any unknown parameters, and are completely determined by the mean composition of the alloy which is assumed to be available.

In the investigated cases, it is justified to assume that the values of the width of the line x_{3j} are identical for different j. Similar simplifications on the basis of physical considerations are used quite frequently. It should be mentioned that the expressions, describing the form of the external peaks of the Mössbauer spectra of the ferromagnetic alloys on the basis of iron in different approximations have the form identical to equation (2.7).

The form of the Mössbauer spectrum of a multiphase alloy may be represented, as in the case of (2.5), in the following form:

$$I(v)=x_1\left[1-\sum_{l=1}^{G}x_2^{(l)}\sum_{j_l=1}^{L_l}P\left(j_l;x_3^{(j_l)},\ldots x_{s_l}^{(j_l)}\right)x_{s_l+1}^{(j_l)}F(x)\left(v;x_{s_l+2}^{(j_l)},\ldots x_p^{(j_l)}\right)\right], \quad (2.8)$$

where the parameters $x_2^{(l)}$ coincide with the previously introduced parameters γ_1, whose $G-1$ independent values are usually not available prior to experiments; $x_{s_l+1}^{(j_l)}=\frac{\pi}{2}\chi n\sigma_0 f_{j_l}^{'}$ (in the approximation of a constant value of f' for different positions – within the limits of the phases – this parameter depends only on l: $x_{s_l+1}^{(l)}=\frac{\pi}{2}\chi n\sigma_0 f_1^{'}$). The meaning of the remaining parameters is the same as in the case of a single-phase alloy. At G = 1, equation (2.8) coincides with equation (2.5).

It should be mentioned that the given complicated enumeration of the unknown parameters, included in equations (2.5) and (2.8), in solving the problem of their determination, is often conveniently replaced by the single-index enumeration and the ordered sets x_k are considered as the vectors of the parameters $\mathbf{x} = \{x_1, x_2,..., x_n\}$ in the n-dimensional space.

Continuous description

The examined case of discrete description took into account the presence of non-equivalent groups of resonant nuclei in the crystal. At the same time, in some cases it is not possible to separate the discrete set of the positions (the spectrum of states) realised with probabilities P_j. The latter is characteristic of complex multicomponent systems, in particular, of amorphous alloys, some phases of variable composition, etc. In addition to this, in some cases because of the low resolution of a large number of components of the 'sub-spectra' it is not possible to determine the unambiguous correspondence between the values of P_j and parameters $x_s^{(j)}$ (or P_{j_l} and $x_{s_l}^{(j_l)}$). In similar cases, there is applied continuous description based, like discrete description, on the linear superposition of the continuously distributed 'sub-spectra'.

Thus, if the difference of the continuously distributed components of the resultant γ-resonance spectrum is determined mainly by one parameter[1] x which is characterised by the

[1]Weak dependence on other parameters is usually ignored because the introduction of multi-dimensional densities is associated with considerable difficulties.

probability density $f(x)$, then:

$$I(v) = x_1 \left[1 - \int_{x_{min}}^{x_{max}} f(x) F(v,x) dx \right]. \tag{2.9}$$

The function $F(v,x)$ specifies the form of the continuously distributed components. In practice, continuous description is sometimes replaced by a quasicontinuous one and, in this case, the integral degenerates into a sum with a specific step. The methods of determination of $f(x)$ are described in the Appendices 1 and 2.

It should be mentioned that in the case of the low resolution of the lines, corresponding to different non-equivalent positions of the resonant atoms, their superposition is often approximated by a broadened line, whose parameters, in the final analysis, contain the averaged-out information on hyperfine interactions in the investigated solid.

Chapter 3

ELECTRICAL AND MAGNETIC HYPERFINE INTERACTIONS OF RESONANT NUCLEI IN METALS AND ALLOYS

3.1. GENERAL CONSIDERATIONS

The parameters δ and H (isomeric shift and the effective magnetic field on the nucleus of the atom of the resonant isotope) reflect the distribution of the charge and spin density of the electrons in the crystal and also the presence of external (applied to the investigated object from the outside) electrical and magnetic fields.

In the majority of cases, the alloys of metals and metallic compounds are characterised by a high degree of symmetry of the crystal lattice. If the symmetry of the lattice is not lower than cubic, the gradient of the electrical field in the area of the nucleus is equal to zero. Consequently, there is no electrical quadrupole splitting of nuclear levels.[1] In this connection, special attention will be given to the electrical monopole and magnetic dipole interaction of the nuclei with bonded and collectivised electrons (and to the respective problems of the formation of the charge and spin density in metallic alloys).

Isomeric shift

Isomeric shift is proportional to the resultant density of the

[1]This does not relate of the interstitial alloys and also to alloys with the anomalously high concentration of point defects. The local distortions of the symmetry and gradients of the electrical field, although they are relatively small in the substitutional alloys, may be determined by the difference of the atomic radii and the charge state of the components. The quadrupole splitting, causing, in the latter case, only broadening of the lines, may be taken into account, for example, using the model of point charges.

electronic charge at the location of the nucleus of the Mössbauer isotope. The formation of chemical compounds is accompanied by changes in the electron density of only the outer valency shells. Similarly, in the formation of alloys, changes are observed only in the outer incompletely filled shells of the atoms. For certain Mössbauer isotopes (in particular, for ^{57}Fe and ^{119}Sn) quantum-chemical calculations were carried out to determine the dependence of the extent of the isomeric shift on the degree of filling of the outer *s* and *d*-electron shells.

Figure 3.1 shows the results of such calculations (using the Hartree–Fock method) for the ^{57}Fe isotope [26]. Since the difference of the radii of the ground and excited states of the nucleus of ^{57}Fe is negative, the isomeric shift increases with a decrease of the density of outer 4*s*-electrons (and vice versa). The

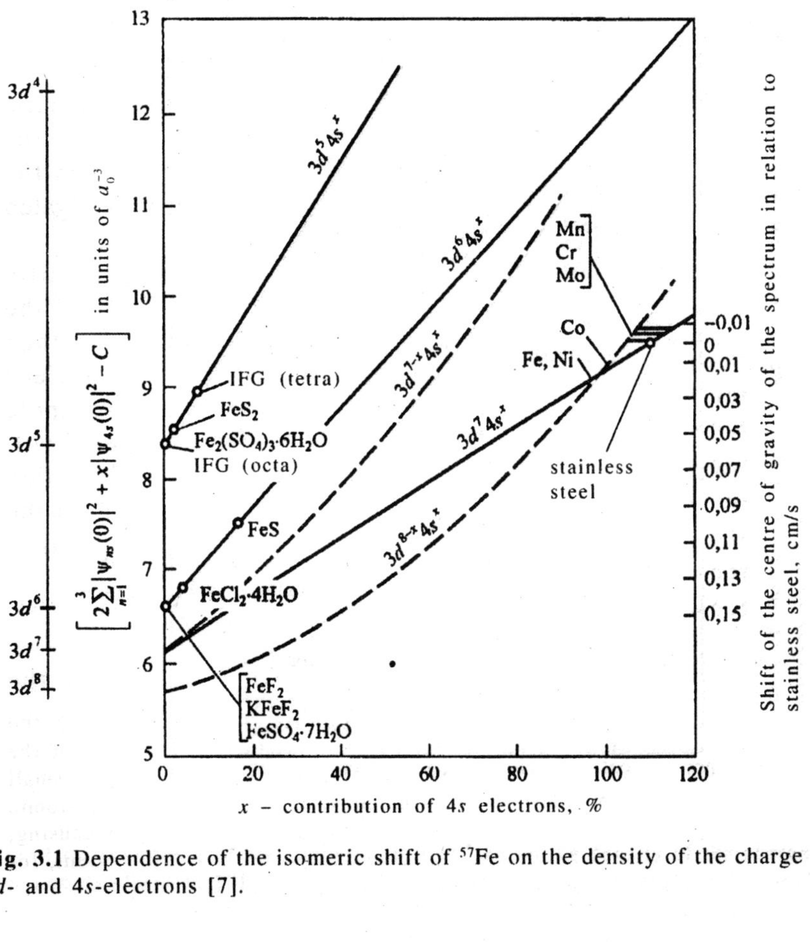

Fig. 3.1 Dependence of the isomeric shift of ^{57}Fe on the density of the charge of 3*d*- and 4*s*-electrons [7].

dependence on the number of $3d$-electrons, screening $4s$-electrons, is of the opposite type.

On the basis of these calculations it has been established that the atoms of iron in pure iron are characterised by the electronic configuration $3d^7 4s^1$ [7].

Effective magnetic field

Initial experiments on the measurement of the effective magnetic fields on the nuclei of the atoms using the Mössbauer effect have stimulated theoretical and experimental investigations in greatly differing areas of the physics of the magnetic state. For example, in the study by Hanna *et al* [15] it was found that the magnetic field on the nuclei of the atoms in pure iron is opposite to the field of the magnetic domain. This contradicted theoretical predictions. The obtained result was explained in a study by Watson and Freeman [27]. They too into consideration the contact polarisation of internal completely filled s-shells. The results of examination of a large number of other systems also required improvement of the accuracy of the available theoretical models.

The corrected knowledge of the mechanisms of formation of effective magnetic fields on the nuclei of the atoms of the magnetically ordered crystals, including metallic alloys, have been generalised in a number of scientific articles and monographs [7, 27–29,31].

One of the main contributions is the contribution of the contact Fermi interaction (contact polarisation of the s-electrons by the resultant spin of the atom[1]) [7, 12]:

$$H_c = \frac{16\pi}{3}\mu_B \left\langle \sum \left(|\psi_\uparrow(0)|^2 - |\psi_\downarrow(0)|^2 \right) \right\rangle, \tag{3.1}$$

where $|\psi_\uparrow(0)|^2$ and $|\psi_\downarrow(0)|^2$ are the spin densities of the s-electrons 'on the nucleus' (i.e. in the location of the nucleus) with the spin directed upwards and downwards, μ_B is the Bohr magneton (summation in respect of all s-shells of the atom).

Another contribution is the contribution from the orbital moment of the electrons:

$$\mathbf{H}_L = 2\mu_B \left\langle \frac{1}{r^2} \right\rangle \langle \mathbf{L} \rangle, \tag{3.2}$$

[1]The resultant spin of $3d$-electrons for iron.

which exists only if the component L_z is characterised by the mean value differing from zero. In splitting of the levels of the ion in the electrical crystal field, the mean value of each component of the orbital moment in each orbital state ψ_i is equal to zero and, in particular, $\langle\psi_i|L_z|\psi_i\rangle=0$. In such a case, it is concluded that the orbital moment is frozen.

Besides, there is a dipole contribution from the spin of the intrinsic atom:

$$\mathbf{H}_\mathrm{D}=2\mu_\mathrm{B}\left\langle 3\mathbf{r}(\mathbf{Sr})\frac{1}{r^5}-\mathbf{S}\frac{1}{r^3}\right\rangle. \tag{3.3}$$

For a cubic crystal, consisting of identical atoms, in the absence of spin-orbital interaction this field is equal to zero.

In the application of the external field $\mathbf{H}_0$ to the crystal, the so-called internal field forms:

$$\mathbf{H}_i=\mathbf{H}_0+\frac{4\pi}{3}\mathbf{M}-\mathbf{H}, \tag{3.4}$$

where $\frac{4\pi}{3}\mathbf{M}$ is the Lorentz field, $\mathbf{H}$ is the demagnetising field induced by the external field.

In order to carry out a more detailed analysis of the effective magnetic fields on the nuclei of the atoms in magnetically ordered crystals it is necessary, in addition to considering the role of intrinsic electron shells of the resonant atom, to take into account the contribution of the magnetic moments of the surrounding atoms and, in addition to this, the effects of polarisation and repopulation of conduction electrons [29–31]. These problems are examined in the paragraph 3.3 of the present chapter.

However, initially it is reasonable to examine the experimental data used as a basis for understanding the relationships governing the perturbation of the charge and spin density of the electrons in ferromagnetic impurity crystals on the basis of 3*d*-metals and, in particular, based on iron, with a suitable Mössbauer isotope ^{57}Fe.

The diluted solid solutions as the phases of variable composition are most simple for analysis. At the same time, many of the relationships typical of these solutions can be transferred to concentrated solid solutions.

Examination of the fine structure of the Mössbauer spectra of the diluted alloys of iron with transition and non-transition elements has greatly facilitated understanding the nature of the effective magnetic field in pure iron.

3.2. EXPERIMENTAL DATA ON THE PERTURBATION OF THE CHARGE AND SPIN DENSITY IN CRYSTALS OF TRANSITION METALS IN DISSOLUTION OF IMPURITIES

The initial investigations into this effect using Mössbauer spectroscopy are represented by the classic studies of M.B. Stearns, S.S. Wilson, A.V. Overhauser [32–34], and J.K. Wertheim, V. Jaccarino, J.K. Wernik and D.N.E. Buchanan [35], carried out at the beginning of the 60s of the 20th century. The studies of the above authors have covered a wide range of different impurities of the 3*d*-, 4*d*- and 5*d*-periods and mutually supplement each other, anticipating the main elements of current representations.

For example, in studies by Stearns *et al*, the authors carried out calculations of the Mössbauer spectra of diluted solid solutions[1] of aluminium, silicon, manganese and vanadium in iron in a model taking into account the interaction in the radius of six coordination spheres in the body-centred cubic lattice of iron. This model supposes $9 \times 7 \times 13 \times 25 \times 9 \times 7$ different possible configurations of filling six nearest coordination spheres of the iron atom by the atoms of the impurity, where 8, 6, 12,..., 6 are the corresponding coordination numbers in the BCC lattice. In order to explain the relationship of this model with the structure of the curves of resonance absorption, we might return to paragraph 1.5 which gave the qualitative description of the Mössbauer spectrum of a diluted iron-based solid solution based on iron in the model of the single coordination sphere.

The authors of the above studies assumed that the intrinsic magnetic moments of the iron atoms, which have atoms of the impurity as their neighbours, remain pratically constant. Because of this, the polarisation of the internal electrons by the intrinsic moments of the iron atoms remains constant, and all changes of the magnetic field on their nuclei are determined by changes of the polarisation of the spin density of conduction electrons, because of the presence of the dissolved impurities close to iron. In fact, it

[1]At the content of substitutional impurities smaller than 10 at% the number of pair impurity–impurity interactions does not exceed 1%.

may be assumed that the formation of the solid solution takes place through removal, from the individual sites, of the iron atoms with the magnetic moment $\mu_{Fe} = 2.2\ \mu_B$ and by placing in their positions the impurity atoms with the moment μ_x. In this case, $\Delta\mu = -\mu_{Fe} + \mu_x$ may have the positive sign, the negative sign or it may be equal to zero. In the case of low absolute values of μ_x (as compared with μ_{Fe}) the value of $\Delta\mu$ for magnetic and non-magnetic impurities is approximately the same, which also explains the approximately same effect [32–34] of the investigated impurities on the value of H_{ef} (see paragraph 3.3.5, equation (3.36), Tables 3.1 and 3.2).

The results of detailed calculations [32–34] show that in diluted alloys, the atoms of the impurity have an additive effect on H_{ef}:

$$\Delta H = \sum_{i=1}^{6} m_i \Delta H_i, \tag{3.5}$$

where m_i is the number of the nodes of the i-th coordination sphere, occupied by the atoms of the impurity and ΔH_i is the contribution to the effective magnetic field from the atom of the impurity in the i-th sphere. The authors concluded that to obtain reliable values of ΔH_1 and ΔH_2 it is necessary to take into account at least the first four coordination spheres of the iron atom.

Figure 3.2 shows the radial dependence of the spin density of the conduction electrons for alloys of iron with aluminium, manganese and vanadium.

Initially, the perturbed spin density was explained by the authors of the study using the Ruderman–Kittel–Kasuya–Iosida (RKKI) theory (see [36] and the relationship (3.20)). However, it was found that the perturbations predicted by this theory are approximately 7 times smaller than those detected in the experiments. Nevertheless, Stearns and Overhauser obtained a good agreement with the experiment [32–34] as a result of introducing a special type of the dependence of spin susceptibility $\chi(q)$ on the wave vector q which has a sharp maximum at $q = 2k_F$. The resultant form of the function $\chi(q)$ was predicted by Overhauser.

In another classic study of Wertheim *et al* [35], attention was given to analysis of the nature of exchange interaction and the distribution of the charge and spin density in iron alloys with different impurities: Ni, V, Cr, Mn, Co, Ru. Al, Ga and Sn (from 4 to 16 at%). Special attention was given to the effect of the nearest and next nearest atoms of the impurity (the two nearest coordination spheres). The authors consider the nodes of the two nearest coordination spheres as unique, owing to the fact that the

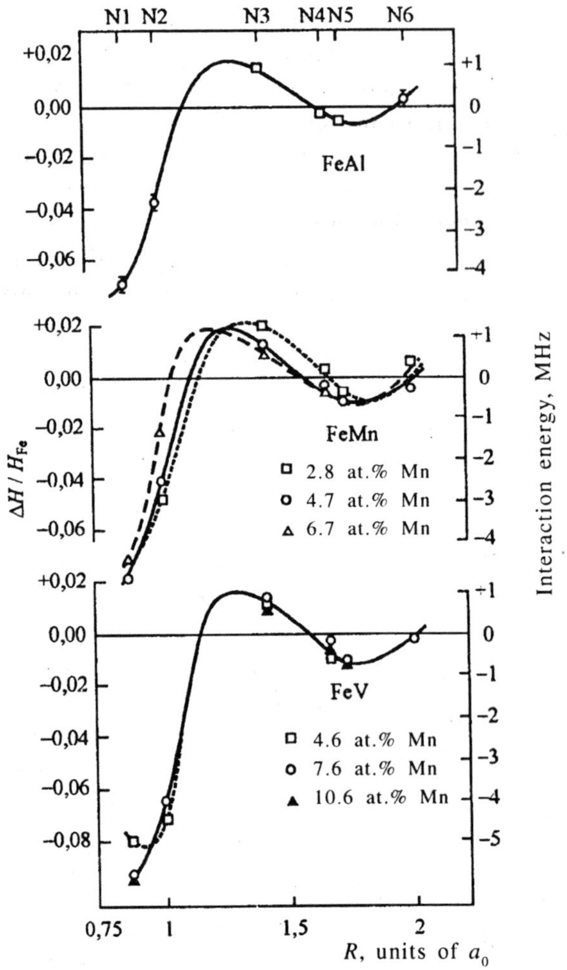

Fig. 3.2 Dependences of the spin density of conduction electrons in FeAl, FeMn and FeV alloys on the distance from the impurity atom, obtained from the shifts of the internal magnetic field. Graphs start at $R = 0.75a_0$, and not with the position of the impurity atom [32].

exchange potential is determined by the direct overlapping of the wave functions of the adjacent atoms (i.e., it is necessary to take into account mainly these exchange interactions). As in the studies by Stearns *et al*, the results show the fact of the additive effect of the impurity atoms on H_{ef} in accordance with the relationship (3.7).

The intensities of the components of the spectrum in the given study were assumed to be proportional to the probabilities of the iron atoms being surrounded by different numbers of the nearest and next nearest atoms of the impurity. The probabilities $P(m, n)$ of the presence of m and n atoms of the impurity in the first and second coordination spheres were defined by the product of two binomial distributions. The external peaks of the Mössbauer spectra were processed. This gave contributions to the effective magnetic field and the isomeric shift. The form of the external peaks $-1/2 \rightarrow -3/2$, $+1/2 \rightarrow +3/2$ was fitted by the function of the type:

$$\varepsilon(E)=\sum_{n=0}^{8}\sum_{m=0}^{6}\frac{8!}{(8-n)!n!}\frac{6!}{(6-m)!m!}\frac{c^{n+m}(1-c)^{14-n-m}}{1+(E-\alpha n-\beta m)^{2}}, \tag{3.6}$$

where α and β have the meaning of $\Delta H_1/2$ and $\Delta H_2/2$ (with the correction for the contribution from the variation of charge density, see sections 3.4 and 4.7, c is the concentration of the impurity in atomic fractions (8 and 6 are the coordination numbers for the first two coordination spheres of the BCC lattice), energy E is measured in units of the width of the resonant line.

On the whole, the resultant values of ΔH_1 and ΔH_2 (Table 3.1) are in agreement with the data published in [32, 33]. The following relationship was obtain for the effective magnetic field:

$$H(m,n)=H_{\text{Fe}}(1+kc)(1+h_1 n+h_2 m). \tag{3.7}$$

where H_{Fe} is the field in pure iron ($h_1 = \Delta H_1/\text{H}_{\text{Fe}}$, $h_2 = \Delta H_2/H_{\text{Fe}}$).

According to the conclusions of the authors, the effect of the impurity on the hyperfine structure of the Mössbauer spectra shows the following relationships.

1. The values of the contributions ΔH_1 and ΔH_2, for the elements positioned to the left of iron in the periodic table of elements, are negative and do not depend strongly on the nature of the impurity.
2. The contributions ΔH_1 and ΔH_2 from the cobalt atoms are of the opposite sign (as well as those of other elements, such as Rh, Ru, Pd, Pt, positioned to the right of iron in the periodic table of elements).
3. The contributions from the nearest and next nearest neighbours of the iron atom are additive.
4. There is no strict regularity in the effect of the impurities on the isomeric shift.
5. The effective magnetic field H_0 on the iron atoms, which do not

Table 3.1 Contributions h_1 and h_2 to the field of hyperfine interaction $H(m,n) = H_{Fe}(1 + kc) \times (1 + nh_1 + mh_2)$ on the nucleus of the Fe atom, with n nearest and m next nearest impurity atoms, field amplification factor k and the variation of the isomeric shift δ (cm/s) per one nearest atom of the impurity [35]

	Al	Si*	Ti	V	Cr	Mn	Fe	Co	Ga	Ru	Sn
h_1	-0.081_5	-0.083	-0.065_5	-0.076_5	-0.083_5	-0.068_5		0.043_5	-0.074_5	-0.067_5	-0.073_5
h_2	-0.046_5	...	-0.06_1	-0.064_5	-0.07_1	-0.01_1		0.021_5	-0.05_1	-0.025_5	-0.02_1
k	0.00_5	0.00	0.05_5	0.3_1	0.4_1	0.1_1		0.04_2	0.00_5	0.2_1	0.06_3
δ	0.000_1	0.004	0.000_1	-0.006_1	-0.007_1	0.000_1		0.000_1	0.008_1	0.000_1	0.004_1

Comment. Multiplier $(1+kc)$, which depends on concentration, may include several effects associated with more distant neighbours. The estimated indeterminacies in the last significant digit are shown in the form of an index.

have any atoms of the impurity in their nearest and next nearest neighbours, has a tendency to increase in comparison with H_{ef} in pure iron.

The authors of the cited study (in contrast to the authors of [32–34]) links the contributions to H_{ef} mainly with the variation of the z-projection of intrinsic magnetic moments of the iron atoms as a result of the change in the local exchange potential[1] (which decreases for all investigated elements, with the exception of cobalt).

On the basis of analysis of the structure of the Mössbauer spectra of iron alloys with the impurities Sawer and Reynik [37] obtained the following equation for the effective magnetic field on the nuclei of the iron atoms depending on the number of the nearest atoms of the impurity and concentration:

$$H(l_1, l_2, \ldots, l_n) = H' + \sum_{i=1}^{n} l_i \Delta H_i, \quad (3.8)$$

where $H' = H_{Fe} + k'c$. The relationships (3.7) and (3.8) differ only slightly

[1]In reality, an important role is played by both the redistribution of the spin density of the conduction electrons and by the variation of the intrinsic moment of the iron atom, see section 3.3.

from each other, taking into account the fact that the coefficients k and k' give only a weak dependence of the field on the concentration, the more so that diluted solutions are considered. The equation (3.37) of a similar form was obtained as a result of the analysis of different mechanisms of the formation of the effective magnetic field on the nuclei of the atoms of the ferro-magnetic matrix (paragraph 3.3).

The experimental fact of the increase of H_{ef} on the nuclei of the iron atoms [35, 37] with no nearest atoms of the impurity, taken into account by the empirical coefficients k and k', is explained in paragraph 3.3 by the effect of repopulation of the conduction electrons (see equations (3.18), (3.35)), and also by the fact that the contribution of the remote coordination spheres is not taken into account.

As shown later, the partial contributions ΔH_i are determined, on the one side, by the variation of the intrinsic magnetic moment of the atoms of the matrix (Fe) (3.21) and, on the other side, by the variation of polarisation (the amplitude of radial oscillations of spin density) of the conduction electrons as a result of the presence of the impurities (see equation (3.20) and figures 3.2 and 3.4).

According to the data in [32–35,37], the majority of the elements, positioned to the left of iron in the periodic table of elements, provide, at least in the first sphere, closely spaced contributions ΔH_i to H_{ef} which, as already mentioned, may be associated with the low value of μ_x relative to μ_{Fe}.

The previously mentioned investigations carried out by Wertheim *et al* [35] and also Campbell and Vincze showed that the contributions to H_{ef} from the nearest atoms of the impurity, situated in the periodic table of elements to the right of iron, have the opposite sign. Their absolute value is smaller. At that, the dependence of the mean field on the concentration for the solid solutions of Co and Ni in iron [38] is nonlinear.

3.3. FERROMAGNETIC IMPURITY CRYSTALS BASED ON TRANSITION METALS

In the present section, the problem is investigated of the role of different mechanisms in the formation of the field of hyperfine interaction on the nuclei of the atoms of the matrix in ferromagnetically ordered crystals of the transition metals with small additions of different impurities. The aim of this investigation is to obtain an analytical relationship of the mean value of the effective magnetic field <H> with the values of the individual atomic magnetic moments on the crystal lattice sites. The latter, together with the well-known Marshall equation for the mean value of the magnetic moment <μ> of the alloy, represents a new independent possibility of determining the individual atomic magnetic moments in the crystals of transition metals. Such an analysis was carried out in [30, 31] and used as the basis of the sections 3.3.1–3.3.5 of the present chapter.

The problem of the behaviour of magnetic and non-magnetic impurities in magnetically ordered crystals of transition metals occupies a very special position in the physics of magnetic phenomena. Until recently, the only generally recognised method of evaluating the values of the magnetic moments[1] of the atoms of the impurity and the matrix was the method of magnetic diffusion scattering of neutrons [40–45]. Evidently, it is therefore important to justify the independent methods of determination of the individual atomic magnetic moments in the crystals, taking also into account the fact that neutron diffraction data, because of a large number of reasons, are characterised by significant errors and require independent confirmation and refinement.

The authors of [39] proposed an equation linking the mean value of the magnetic moment and the effective magnetic field ($H_{ef} \equiv H$) on the nuclei of the iron atoms in the diluted alloys of iron with different impurities:

$$\partial\langle H\rangle/\partial c = A\left(\partial\langle\mu\rangle/\partial c + \mu_{\mathrm{Fe}} - \mu_x\right) + B\left(-\mu_{\mathrm{Fe}} + \mu_x\right), \tag{3.9}$$

where μ_x is the value of the z-projection of the magnetic moment on the impurity node; μ_{Fe} is the magnetic moment of pure iron. This relationship is heuristic because it is derived on the basis of the most general qualitative considerations. Consequently, the relationship between the <H> and <μ> and the validity of the equation (3.9) require more detailed theoretical analysis. It is necessary to evaluate the role of some of the mechanisms of formation of the field, such as RKKI polarisation and repopulation of conduction electrons, not examined in [39], and it is also important to take into account the lattice dipole contributions to H_{ef}.

The values of H_{ef} on the nuclei of the atoms in magnetically ordered alloys may be reliably measured by nuclear gamma resosnance and nuclear magnetic resonance methods. The validation

[1]Here and further, z-projections of magnetic moments are meant.

of the equations, describing the relationship of the values of H_{ef} with the value of the nodal magnetic moments, opens possibilities for the experimental determination of the latter, and also for investigating the dependence of the magnitude of these moments on the local environment. In this case, it is useful to involve the data on magnetisation. It should be mentioned that regardless of the existence of a certain correlation in the effect of the impurities on $<\mu>$ and $<H>$, no complete analogy was detected for the concentration dependences of these quantities (Fig. 3.3 *a*, *b*), and this also requires explanation.

3.3.1. Analysis of the role of different mechanisms in the formation of H_{ef} on atom nuclei

The views on the role of different mechanisms in the formation of the field of the hyperfine interaction on iron nuclei are rather contradicting. In the available studies, preference is usually given to individual contributions (different – by different authors). It is necessary to examine this, in general, non-trivial problem in greater detail, with the aim of obtaining, in the final analysis, the relationship of H_{ef} with the values of the nodal moments.

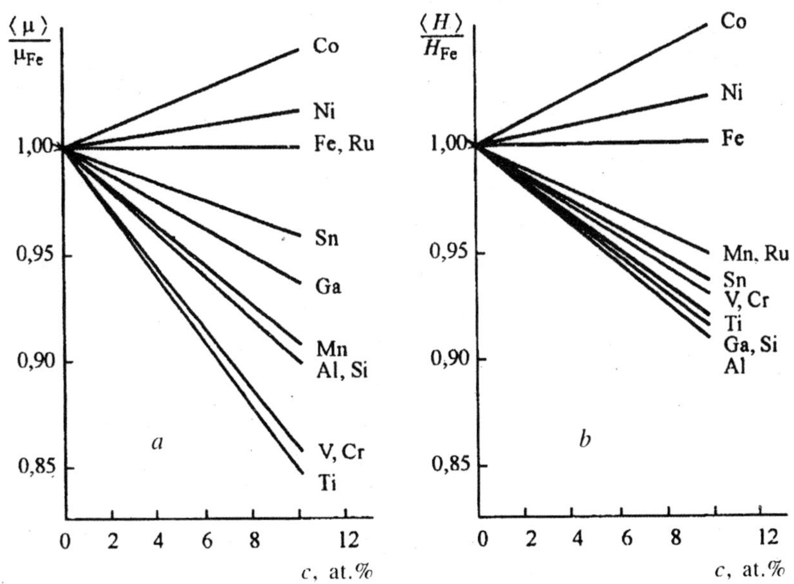

Fig. 3.3 Concentration dependences of the relative values of $<\mu>$ [49] and $<H>$ [35] for various impurities.

The strength of the field in pure iron

Analysis of the currently available experimental data shows that the effective magnetic field on the nuclei of the iron atoms in pure iron in the absence of external fields may be represented [20,29] in the form of the sum of the following main contributions[1]:

$$H = H_{(1,2,3)s} + H_{4s} + H^*, \tag{3.10}$$

where $H_{(1,2,3)s}$ is the contact field determined by the polarisation of the electrons of the ion core of the iron atom by the intrinsic magnetic moment (intrinsic 3*d*-electrons); H_{4s} is the contribution of the conduction electrons, polarised by the intrinsic moment of Fe; $H^* = H^*_{(1)} + H^*_{(2)}$ is the contribution from the electrons of the conduction band, determined by two main effects: 1. Oscillating polarisation of spin density, and 2. Homogeneous (over the crystal) repopulation (Zener effect) of the conduction electrons as a result of the exchange interaction with the localised moments of the impurity atoms.

The equation (3.10) does not contain the dipole contribution H_D from the magnetic moments of the adjacent atoms. In pure iron, because of the cubic symmetry of the lattice, the resultant dipole contribution from the magnetic ions of Fe, distributed around the central iron atom, is equal to zero (see the relationship (3.13)).

The separation of the first two contributions in (3.10) is extremely complicated. In this connection, the linear dependence of the sum of these contributions on the value of the intrinsic magnetic moment of the iron atom determined on the basis of the analysis of a large number of experimental data is especially favourable (see for example [29]):

$$H = \alpha\mu_{Fe} + H^*_{(1)} + H^*_{(2)}, \tag{3.11}$$

where μ_{Fe} is the *z*-projection of the magnetic moment of the iron atom in the alloy. In the first approximation, the linear form of the dependence of the field of the contact interaction on μ_{Fe} may be explained by the fact that the highest measured values of H_{ef} correspond to the contact polarisation of *s*-electrons of only several percent. Since the variation of μ_{Fe}, induced by the atoms of the

[1]All analysed contributions are either parallel or antiparallel to the magnetisation axis; therefore, equation (3.10) permits the scalar form to be used.

impurity, in the diluted alloys is relatively small, it will be assumed that at fixed temperature α = const and the investigated linear approximation is valid.

Effect of impurity atoms

The appearance of impurity atoms in the vicinity of the iron atoms leads to a change of the value of H as a result of a change of the values of μ_{Fe}, H^* and the emergence of the lattice dipole contribution ΔH_D, formed at distortion of the magnetic symmetry of the environment of the iron atom:

$$\Delta H = \alpha \Delta \mu_{Fe} + \Delta H^*_{(1)} + \Delta H^*_{(2)} + \Delta H_D. \tag{3.12}$$

(It should be mentioned that the approximation $H = H_{Fe} + \alpha\Delta\mu_{Fe}$ in the vicinity of $H = H_{Fe}$ with the appropriate selection of α is more realistic than $H_{Fe} = \alpha\mu_{Fe}$).

We examine the nature of the individual contributions.

1. The variation of the moment of the central atom ($\Delta\mu_{Fe}$). This effect is determined by the overlapping and covalent mixing of the bounde electronic state of iron and of the impurity atoms. The different signs of the addends, formed as a result of overlapping and covalence complicate the accurate prediction of the resultant total spin. It is well-known that the value of H_{ef} decreases with an increase of the covalent component of the interaction. Of obvious interest is the possibility of the experimental determination of the value of μ_{Fe}.

2. The lattice dipole contribution (ΔH_D). Because pure iron is characterised by the total vector compensation of the lattice dipole contributions to H_{ef}, the disappearance of the magnetic ion of Fe in substitution by a non-magnetic impurity may formally be regarded as the formation, on the lattice node, of the negative moment $\Delta\mu = -\mu_{Fe}$. If the impurity is characterised by intrinsic magnetic moment μ_x, then $\Delta\mu = -\mu_{Fe} + \mu_x$ (for co-linear ferromagnetics the scalar form of presentation is justified). The respective variation of the effective magnetic field on the nucleus is determined by the following relationship:

$$\delta \mathbf{H}_D = \frac{3(\Delta\boldsymbol{\mu}\mathbf{n})\mathbf{n} - \Delta\boldsymbol{\mu}}{R^3}, \quad \mathbf{n} = \frac{\mathbf{R}}{R} \tag{3.13}$$

(R is the distance between the impurity atom and the iron atom), which shows that for BCC iron-based alloys the dipole contribution to the effective field from the impurity atom in the first coordination sphere is small in the absolute value ($<0.01\ H_{Fe}$) and is perpendicular to the z-axis. Consequently, for any configuration of the nearest environment of the iron atom by the impurity atoms, the absolute value of the addition to H_{ef} is either equal to zero (because of the vector compensation of the contributions from the individual atoms) or it is negligible. Therefore, the main (not negligible) contribution to ΔH_D in the BCC lattice is determined by the effect of the second coordination sphere and also depends on the number and mutual position of the impurity atoms in the second sphere. In this case, the contribution is either parallel or anti-parallel to the axis z, which permits inclusion of ΔH_D in (3.12) in the scalar form. For ΔH_D from (3.13) we obtain:

$$\Delta H_D \approx \Delta H_D^{(2)} = -\frac{\Delta\mu}{R_{(2)}^3} F_{(2)}, \tag{3.14}$$

where $F_{(2)} = \sum_{i=1}^{l_2}(3\cos^2\theta_i - 1)$ is the structural factor; θ_i is the angle between $\boldsymbol{\Delta\mu}$ and $\mathbf{R}_{(2),i}(|\mathbf{R}_{(2),i}|=R_{(2)})$; l_2 is the number of the impurity atoms in the second coordination sphere of the iron atom; i is the index numerating different positions of the impurity atom in the second coordination sphere of the iron atom. The minus sign in equation (3.14) takes into account the fact that the magnetic field on the nuclei of the iron atoms is anti-parallel in relation to the direction of magnetisation.

3. Effects of polarisation and repopulation of the conduction electrons (ΔH_1^*, ΔH_2^*). According to the current views, in the dissolution of 'non-magnetic' impurities (of the type of aluminium and silicon) in iron whose small amounts have almost no effect on the magnetic and zone structure of the matrix, the main contribution to H^* is provided by the oscillations of the spin density of the conduction electrons [20, 24].

In RKKI theory, using the Hamiltonian of the interaction of the individual local moment with a conduction electron of the type:

$$H = \sum_j A(\mathbf{x}_j - \mathbf{r}_n)\boldsymbol{\sigma}_j \mathbf{S}_n, \tag{3.15}$$

where $\boldsymbol{\sigma}_j$ and $\mathbf{S}_n$ are the electronic and atomic spins; $\mathbf{r}_n$ and $\mathbf{x}_j$ are the coordinates of the localised moment and the electron; $A(\mathbf{x}_j-\mathbf{r}_n)=$

$J\delta(\mathbf{x}_j-\mathbf{r}_n)$ is the integral of the $\{s-d(f)\}$-exchange, assuming the contact nature of the exchange of the local moment with the conduction electron, we obtain the following analytical expressions for the magnitude of perturbation of the spin density of the conduction electrons at the distance R from an impurity and the corresponding contribution $\delta H^*_{(1)}$ to the effective field [47]:

$$\Delta\rho(R) = -\Delta\mu \frac{18\pi n_e^2}{gE_F} JF(2k_F R), \tag{3.16}$$

$$\delta H^*_{(1)} = A_{4s}(0)\Delta\rho(R) \equiv b_{(1)}\Delta\mu F(2k_F R), \tag{3.17}$$

where $\Delta\mu = -\mu_{Fe} + \mu_x$ in the units of μ_B; g is the Lande factor; $2n_e$ is the number of the $4s$-conduction electrons per atom; E_F is Fermi energy; $F(x)$ is the Ruderman–Kittel function: $F(x) = (x \cos x - \sin x)/x^4$; k_F is the pulse of the electron on the Fermi surface; $A_{4s}(0)$ is the contribution of the single s-conduction electron to the magnetic field on the nucleus in the units of kOe/μ_B. It is assumed that the constants of the $\{s\text{-}d(f)\}$-exchange do not depend on the nature of the impurity.

The results of calculation of $\Delta\rho(R)$ using equation (3.17) are in qualitative agreement with the results of the Mössbauer determination of the variation of spin density on the iron nuclei caused by the non-magnetic atoms of aluminium and silicon of several nearest coordination spheres [20]. Another assumption on the form of the integral of the $\{s-d(f)\}$-exchange [36] also leads to the oscillating nature of the dependence $\Delta\rho(R)$ and ensures quantitative agreement with the experiments.

The Zener mechanism, taking into account the repopulation of the conduction electrons, results in the perturbation of the spin density from every individual impurity with uniform distribution in the volume of the crystal [28] and also proportional to $\Delta\mu$:

$$\delta H^*_{(2)} \sim \Delta\mu A_{4s}(0)\frac{3n_e J}{gE_F} \equiv b_{(2)}\Delta\mu. \tag{3.18}$$

It should be mentioned that the Zener mechanism assumes the variation of the population of the states with the spin 'upwards' and the spin 'downwards' (i.e., repopulation with inversion of the spin), whereas the RKKI polarisation results in different-phase 'densening' and 'rarefaction' of the spins of different sign without variation of the spin direction.

The resultant values of the contributions $\Delta H^*_{(1)}$ and $\Delta H^*_{(2)}$ to H_{ef} on the nuclei of the iron atoms in the linear approximation can be determined, as shown later, by the summation of the contributions from the individual impurity atoms.

3.3.2. Statistical fluctuations of the local environment

Since in the tendency to zero of the concentration of the impurity the value of k_F should approach $k_F^{(Fe)}$, taking into account the constancy of the constant of the $\{s\text{–}d(f)\}$-exchange for different impurities, the function (3.16) in analysis of diluted solutions may be used as a reference function, independent of the nature of the impurity. Actually, the curves $\Delta\rho(R)$ (3.16) for the impurities with different values of the intrinsic magnetic moment μ_x differ only by the amplitude of the oscillations, proportional to $\Delta\mu = -\mu_{Fe} + \mu_x$ (Fig. 3.4).

The variation of $H^*_{(1)}$ for an arbitrary j-th atom of iron, assuming the independence of the polarisation contributions from the individual impurity atoms may, in accordance with equation (3.16), be determined using the expression:

$$\Delta H^*_{(1),j} = \frac{18\pi n_e^2}{gE_F} J\, A_{4s}(0) \sum_{i \neq j} \Delta\mu_i F\left(2k_F R_{ij}\right), \tag{3.19}$$

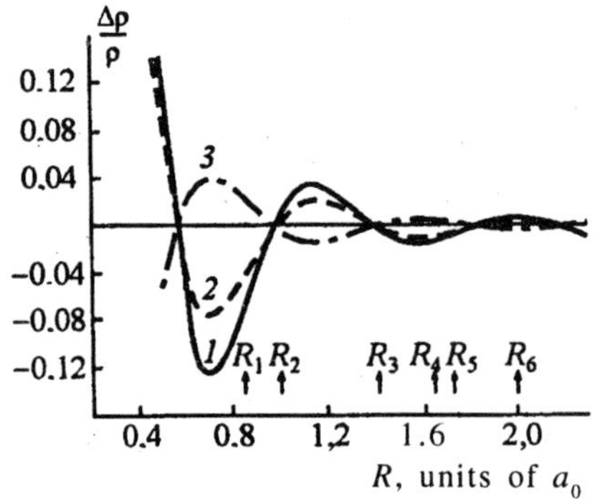

Fig. 3.4 Oscillations of the spin density of conduction electrons for impurities of type *1*, *2* and *3* (curves *1*, *2* and *3*), dissolved in iron $|\Delta\mu_1|>|\Delta\mu_2|>|\Delta\mu_3|$, ($\Delta\mu_1<0$, $\Delta\mu_2<0$, $\Delta\mu_3>0$); $k_F = k_F^{(Fe)} = 3.9\ a_0^{-1}$ (a_0 is the lattice parameter of pure iron, R_i are radii of coordination spheres) [31].

where summation takes place over the nodes occupied by the impurity atoms. At low concentrations of the impurities, this approximation is quite accurate. For a binary alloy, the relationship (3.19) may be re-written in the form of the sum of the contributions from the sequence of n coordination spheres:

$$\Delta H^*_{(1)} = b_{(1)}\Delta\mu \sum_{i=j} l_i F(2k_F R_i), \qquad (3.20)$$

where l_i is the number of the impurity atoms in the i-th coordination sphere of the selected iron atom. However, the variation of the intrinsic magnetic moment of the iron atom because of the presence in the vicinity of the impurity atoms may be represented [27, 45] in the following form:

$$\Delta\mu_{Fe} = \sum_{i=1}^{n} l_i \Delta\mu_{Fe}(R_i), \qquad (3.21)$$

where $\Delta\mu_{Fe}(R_i)$ is the contribution to the variation of the magnetic moment of iron from the atom of the impurity in the i-th coordination sphere.

It is interesting to calculate the mean (in respect of random configurations of the impurities) value of the field of hyperfine interaction $\langle H \rangle$ on the iron nuclei. Symmetry considerations show that in the cubic matrix, the mean values of the lattice dipole contributions $\langle H^{(i)}_D \rangle$ (i = 1, 2,...) from the first, second, and so on coordination spheres and their resultant mean value $\langle \Delta H_D \rangle$ are equal to zero.

Consequently,

$$\langle \Delta H \rangle = \alpha \langle \Delta\mu_{Fe} \rangle + \langle \Delta H^*_{(1)} \rangle + \langle \Delta H^*_{(2)} \rangle. \qquad (3.22)$$

Further, since $\langle l_i \rangle = cz_i$, where c is the concentration of the impurity, and z_i is the number of the nodes in the i-th coordination sphere, in calculation for the single iron atom we obtain:

$$\langle \Delta\mu_{Fe} \rangle = c \sum_{i=1}^{n} z_i \Delta\mu_{Fe}(R_i); \qquad (3.23)$$

$$\langle \Delta H^*_{(1)} \rangle = b_{(1)} \Delta\mu c \sum_{i=1}^{n} z_i F(2k_F R_i); \qquad (3.24)$$

$$\langle \Delta H^*_{(2)} \rangle = \Delta H^*_{(2)} = b_{(2)} \Delta \mu c, \tag{3.25}$$

where $b_{(2)} = 3A_{4s}(0)n_e J/(gE_F)$ (see equation (3.18)). Finally, we obtain:

$$\langle H \rangle = H_{\text{Fe}} + \alpha c \sum_{i=1}^{n} z_i \Delta \mu_{\text{Fe}}(R_i) + b_{(1)} \Delta \mu c \sum_{i=1}^{n} z_i F(2k_F R_i) + b_{(2)} \Delta \mu c, \tag{3.26}$$

where H_{Fe} is the strength of the field in pure iron. Taking into account that at $c \to 0$ the value $k_F \to k_F^{\text{Fe}}$ (because of this, the second sum for diluted alloys may be regarded as a constant, independent of the type of impurity) and denoting in the limit $n \to \infty$: $\beta = b_{(1)} \sum_{i=1}^{\infty} z_i F(2k_F R_i) + b_{(2)}$, we can write the following equation [30, 31]:

$$\langle H \rangle = H_{\text{Fe}} + \alpha c \sum_{i=1}^{\infty} z_i \Delta \mu_{\text{Fe}}(R_i) + \beta c \Delta \mu, \tag{3.27}$$

where α and β are constants, independent of the nature of the impurity.

3.3.3. Method of determination of atomic magnetic moments on the basis of measurements of magnetisation and H_{ef}

The concentration behaviour of the mean magnetic moment in the region of low concentrations of the impurity is described by the Marshall equation [48]:

$$\partial \langle \mu \rangle / \partial c = \sum_{i=1}^{\infty} z_i \Delta \mu_{\text{Fe}}(R_i) + \Delta \mu; \tag{3.28}$$

and differentiating (3.27), we obtain (see [30, 31])

$$\partial \langle H \rangle / \partial c = \alpha \sum_{i=1}^{\infty} z_i \Delta \mu_{\text{Fe}}(R_i) + \beta \Delta \mu, \tag{3.29}$$

and, finally:

$$\mu_x = \mu_{\text{Fe}} + \frac{a \partial \langle \mu \rangle / \partial c - \partial \langle H \rangle / \partial c}{\alpha - \beta}, \tag{3.30}$$

$$\sum_{i=1}^{\infty} z_i \Delta \mu_{\text{Fe}}(R_i) = \frac{\partial \langle H \rangle / \partial c - \beta \partial \langle \mu \rangle / \partial c}{\alpha - \beta}, \tag{3.31}$$

where $\sum_{i=1}^{\infty} z_i \Delta\mu_{Fe}(R_i) = \partial\langle\mu_{Fe}\rangle/\partial c$. The parameters α and β, as indicated by the above examination, are independent of the nature of the dissolved impurity and are determined only by the parameters of the metal-solvent. If the estimates of the constants α and β are available, the resultant equations (3.30) and (3.31) represent in fact the method of determination of atomic magnetic moments on the basis of the results of measurements of magnetisation and H_{ef}. These equations make it possible to compare the data for magnetisation, nuclear gamma resonance, nuclear magnetic resonance and magnetic neutron diffraction, including the results of determination of the individual atomic magnetic moments by these methods (see the following section).

3.3.4. Calculation of nodal magnetic moments for iron alloys with different impurities. Comparison with neutron diffraction data

The values of the parameters for iron alloys (α = 360 kOe/μ_B, β = 100 kOe/μ_B) were determined by approximation (using equation (3.30)) of the results of neutron diffraction determination of the values of μ_x for the impurities of 3*d*-metals in iron [38], using the data on $\partial\langle\mu\rangle/\partial c$ [49] and $\partial\langle H\rangle/\partial c$ [35] (nuclear gamma resonance), obtained at $T \ll T_c$. It is interesting to note that the processing of somewhat different data [39] for the elements of the 3*d*-period, and also 4*d*- and 5*d*-periods gives very similar values of the constants α and β (α = 350 kOe/μ_B, and β = 100 kOe/μB). Approximation of the same data separately for 3*d*-, 4*d*- and 5*d*-periods at the fixed value of β = 100 kOe/μ_B gives the values of α in the range 320÷440 kOe/μ_B. Taking into account the large errors of the initial data and the small representative sample, it may be assumed that the estimates obtained in this manner are very stable. In [30, 31] we used the above data α = 360 kOe/μ_B, β = 100 kOe/μ_B, determined by the approximation of the data obtained in [38] for a number of 3*d*-metals. These values were assumed to be constant for all impurities and used for the processing of the independent results in [39] and some other studies in order to compare the results with the neutron diffraction data.

Since the coefficients α and β are quite different in magnitude, the concentration dependence of the mean magnetic moment differs from the concentration dependence of the mean value of the effective magnetic field. At that, the equations (3.28) and (3.29) are linearly independent and make it possible to calculate the values of local moments. Of course, these equations in the substitution of the calculated values of μ_x and $\sum z_i \Delta\mu_{Fe}(R_i)$ describe accurately the concentration dependences $<\mu>$ and $<H>$ for different impurities (Fig. 3.3 *a*, *b*).

As indicated by Table 3.2, the agreement of the calculated (from equations (3.30) and (3.31)) and neutron diffraction [38, 39] values of μ_x for the elements of the 3*d*-period, taking into account the estimates of the μ_x errors, is completely satisfactory (graphical comparison of these data is presented in [30]) regardless of the very large error of the initial data. In fact, the reliable interpolation data on $\partial<H>/\partial c$ at the limit $c \to 0$, in which the equations (3.28) and (3.29) are also valid, are not available at the moment. It should be mentioned that beside the inaccuracies of the composition, large errors may result from the uncontrollable deviations of the state of the alloys from the ideally disordered one and a number of other factors.

For the elements of the nickel and cobalt subgroups, i.e. Co, Ni, Rh, Ir, Pt, the calculated values of μ_x are positive and definitely exceed the magnetic moment of pure iron.

This difference is larger for the cobalt subgroup and equals on average 0.7 μ_B in comparison with 0.2 μ_B for the nickel subgroup. It should be mentioned that with the signs of μ_x being identical, neutron diffraction results in considerably lower absolute values of the moments, in particular for Ir and Pt. However, our data published in [30, 31] indicate that the elements of these subgroups behave on the whole in the same manner, somewhat increasing the ferromagnetic properties of the solid solution, both as a result of the presence of the moment exceeding μ_{Fe} in the impurity node and because of the increase of the mean moment of the iron atoms.

The clear difference in relation to the neutron diffraction data is observed in the case of Mo and also for W and Re: the obtained values of μ_x, in contrast to the neutron diffraction data, are positive. At the same time, examination shows a large reduction of the mean magnetic moment of iron (as a result of the reduction of the magnetic moments of the iron atoms bordering with the impurity atoms) – this also explains the high absolute values of $\partial<H>/\partial c$ at a relatively moderate rate of decrease of magnetisation. The data for ruthenium, taking into account the large difference in the results obtained by different authors, require clarification.

The values of μ_x for Al and Si are equal to zero in the range of

Table 3.2 Calculated atomic magnetic moments

Impurity	$\partial\langle H\rangle/\partial c$ kOe		$\partial\langle\mu\rangle/\partial c$ μ_B		Calculated results		Neutron diffraction data	
	[35]	[39]	[49]	[39]	$\frac{\partial\langle\mu_{Fe}\rangle}{\partial c}, \mu_B$	μ_x, μ_B	μ_x, μ_B [38]	μ_x, μ_B [39]
Al	–308	–	–2.28	–	–0.3	0.2	–	–
Si	–295	–	–2.28	–	–0.3	0.2	–	–
Ga	–297	–	–1.43	–	–0.6	1.4	–	–
Sn	–214	–	–0.97	–	–0.5	1.7	–	–
Ti	–276	–270	–3.39	–3.40	0.2(0.3)	–1.4(–1.5)	-1.2_5	–0.9
V	–230	–230	–3.28	–3.30	0.4(0.4)	–1.5(–1.5)	-0.9_3	–0.7
Cr	–228	–240	–2.28	–2.40	0.0(0.0)	–0.1(–0.2)	-0.9_3	–0.8
Mn	–168	–150	–2.11	–1.60	0.2(0.0)	–0.1(–0.6)	0.1_5	0.1
Co	170	205	1.00	1.20	0.3(0.3)	2.9(3.1)	2.1_3	1.9
Ni	67*	160	0.33	0.60	0.1(0.4)	2.4(2.4)	1.2_3	0.9
Mo	–	–380	–	–2.10	(–0.7)	(0.8)	–	–0.7
Ru	–162	–90	0.00	–1.50	–0.6(0.2)	2.8(0.5)	0.7_4	0.8
Rh	–	190	–	1.00	(0.4)	(2.9)	–	0.7
W	–	–400	–	–2.00	(–0.8)	(1.0)	–	–0.6
Re	–	–410	–	–1.60	(–1.0)	(1.6)	–	–0.4
Os	–	–240	–	–1.60	(–0.3)	(0.9)	–	0.0
Ir	–	60	–	0.60	(0.0)	(2.8)	–	0.2
Pt	–	560	–	1.70	(1.5)	(2.4)	–	0.1

Comment. The asterisk shows the data from [51]. The values in the brackets are the results calculated using the data published in Ref [39]. The estimated error of the calculated values $\partial <\mu_{Fe}>/\partial c$ and μ_x is $\pm 0.5\mu_B$; the errors of the values of μ_x [38] are given in the form of indices; in [39] the estimates of the errors are not presented.

the determination error. Positive values were obtained for Ga and Sn. These values are in correlation with the number of external electrons of gallium and tin; consequently, it may be concluded that a ferromagnetic bond (interaction) may exist there. The neutron diffraction data [45] also indicate the possibility of 'inleakage' of the spin density in the areas of location of the non-magnetic impurities, dissolved in pure iron. On the whole, although in the case of the elements of 4*d*- and 5*d*-periods there are differences in comparison with the neutron diffraction data, the obtained results are highly logical. In fact, analysis of the equations (3.28) and (3.29) shows that the value of $<\mu>$ reacts to the same extent to the variation of the magnetic moments of the impurity and the matrix, whereas $<H>$ is characterised by a stronger response to the variation of the intrinsic magnetic moment of the Fe atom and by a considerably weaker response (in $\alpha/\beta = 3.6$ times) to the variation of the magnetic moment in the impurity node. Taking this into account, it becomes understandable that the high (in comparison with the majority of other impurities) rates of decrease of $<H>$ at a relatively low rate of variation of $<\mu>$ with the composition, as, for example, in the case with the Mo, W and Re (Table 3.2), are associated primarily with a larger decrease of the magnetic moments of the iron atoms. On the contrary, the anomalously high value of $\partial<\mu>/\partial c$, greatly exceeding 2.2 μ_B, at a simultaneous large change of the field is characteristic of the impurity–matrix antiferromagnetic interaction (Ti, V).

It should be mentioned that the accuracy of determination of the derivatives $\partial<\mu_{Fe}>/\partial c$ is, because of the large error of the initial data, low in the majority of cases in order to be able to interpret with a certain degree of reliability the relationships governing the effect of the different impurities on the mean moment of the iron atom.

Nevertheless, in the presence of reliable experimental data as well as verified estimates of the parameters α and β, the results of measurements of $<\mu>$ and $<H>$ represent a new independent possibility of determining the values of the individual atomic moments in magnetically ordered crystals on the basis of transition metals.

The conducted examination shows that instead of the well-known equation

$$\langle H\rangle = \alpha\mu_{Fe} + b\langle\mu\rangle, \tag{3.32}$$

which in fact has no theoretical substantiation, it is necessary to use the relationship:

$$\langle H\rangle = H_{Fe} + a(\langle\mu\rangle - \mu_{Fe}) + (\beta - \alpha)c\Delta\mu, \tag{3.33}$$

which is a consequence of (3.28) and (3.29).

The parameter $\beta = b_1 \sum_{i=1}^{\infty} z_i F\left(2k_F^{Fe} R_i\right) + b_{(2)}$ can be regarded as constant only in the range of relatively low concentrations. This results in a linear relationship between $<H>$ and c, and also between $\partial<H>/\partial c$ and $\partial<\mu>/\partial c$. With the increase of the concentration of the impurity, parameter β may greatly change (up to the change of the sign) because of the variation of the value of k_F of the alloy. This may result in the nonlinear dependence of $<H>$ on concentration [50]. This pattern was detected, for example, for Fe–Co and Fe–Ni alloys.

It should be mentioned that the results of analysis taking into account the specific mechanisms of polarisation and repopulation of the conduction electrons, and also the presence of lattice dipole contributions to H_{ef}, confirm the validity of the equation (3.9) which may be easily obtained as a consequence of (3.28) and (3.29) (α and β coincide in the meaning with A and B).

3.3.5. Physical meaning of partial contributions to H_{ef}. The equation for the isotropic part of the contribution

As shown, the mean values of the dipole (anisotropic[1]) contributions $<\Delta H_D^{(i)}>$ (i = 1, 2, 3,...) to H_{ef} in the case of the random distribution of the atoms are equal to zero[2]. This indicates in particular that the dipole contributions may lead only to the broadening of the lines of the Mössbauer spectrum but not to the shift of the lines. Then, according to (3.12), (3.20), (3.21) and (3.25), it is possible to write the following equation for the value of the effective magnetic field, averaged out in respect of different values of the dipole contributions (i.e. for the isotropic part of the magnetic field):

$$H(l_1, l_2, ..., l_n) = H_{Fe} + b_{(2)}\Delta\mu c + \sum_{i=1}^{n} l_i \left[\alpha\Delta\mu_{Fe}(R_i) + b_{(1)}\Delta\mu F(2k_F R_i)\right], \quad (3.34)$$

which shows that the value of H_{ef} on the nuclei of the Fe atoms, characterised by the purely 'iron' environment, is described by the

[1]The term is taken from [28]. It should be mentioned that the potential of the isotropic interaction, in contrast to anisotropic one, is spherically symmetric.
[2]Averaging is carried out in respect of different possible variants of the arrangement of l_i atoms of the impurity in the i-th coordination sphere of iron.

relationship:

$$H(0,0,\ldots,0) = H_{\text{Fe}} + b_{(2)}\Delta\mu c. \tag{3.35}$$

Introducing the notation:

$$\Delta H_i = \alpha\Delta\mu_{\text{Fe}}(R_i) + b_{(1)}\Delta\mu F(2k_{\text{F}}R_i), \tag{3.36}$$

gives the following equation

$$H(l_1, l_2, \ldots, l_n) = H(0,0,\ldots,0) + \sum_{i=1}^{n} l_i \Delta H_i, \tag{3.37}$$

which is identical with the equation, obtained in experiments for different impurities carried out in [37]. In the same study, the authors determine the linear form of the dependence $H(l_1, l_2, \ldots, l_n)$ on concentration which is in agreement with (3.34).

As indicated by (3.35), the value of $H(0, 0, \ldots, 0)$ is determined by the effect of repopulation of the conduction electrons. It should be mentioned that for the impurities of the 3*d*-period (see Table 3.2), this contribution should be antiparallel to the magnetisation axis *z* and should result in an increase of H_{ef} which was also found in the experiments [35, 37].

Thus, equations (3.34)–(3.36) provide information on the physical nature of the contributions to H_{ef} and make it possible to express these contributions by means of the values of the local magnetic moments in the nodes of the crystal lattice. It will be shown that, because of the change in the number of charge carriers and of k_{F} in doping with different impurities, the values of contribution ΔH_i may show a strong dependence on concentration which is detected, for example, for Fe–Mn alloys [50, 51].

3.4. HYPERFINE STRUCTURE OF MÖSSBAUER SPECTRA OF DILUTED IRON-BASED SOLID SOLUTIONS

In previous paragraphs of chapter 3 we examined the problems of formation of the charge and spin density of the electrons on the nuclei of the atoms of the transition metal-solvent.

Since the effects of saturation in the case of the charge density are manifested at a higher content of the impurity in comparison with spin density [28], the variation of the isomeric shift should be described by the dependence identical with (3.8), (3.34). This fact has been established by experiments on the basis of Mössbauer measurements [37]. The following equation may be put down:

$$\delta(l_1,l_2,...,l_n)=\delta(0,0,...,0)+\sum_{i=1}^{n}l_i\Delta\delta_i. \tag{3.38}$$

The electrical and magnetic interactions in the solid solutions based on transition metals are described quite accurately by the model of pair interactions and are determined by the number of the atoms of the dissolved impurity in the n nearest coordination spheres of the resonant isotope. The variations of the disposition of the impurity atoms in relation to each other play a considerably smaller role.

As already mentioned in paragraph 3.3.5, the mean value of the anisotropic contribution, associated with magnetic dipole interaction (3.14) for disordered solid solutions, is equal to zero. Consequently, anisotropic interaction results only in the broadening of the Mössbauer lines without causing their shift[1].

Taking into account the results obtained in this chapter, we may describe, in the explicit form, the position of the lines of the hyperfine structure for substitutional iron-based solid solutions with a low content of the impurities.

Figure 1.4 shows the Zeeman sextet of lines of pure iron at T = 300 K. The positions of the lines on the scale of energies, denoted by E_k (k = 1, 2, ..., 6 – with increasing energy), may be described by the relationship:

$$E_k=\delta_{Fe}+A_k H_{Fe}. \tag{3.39}$$

The numbers A_k are determined by the fine structure of the nuclei levels of ^{57}Fe (see equation (1.18) and paragraph 4.4, Chapter 4).

The dissolution of the impurities results in the formation of a wider spectrum of energies. Taking into account (3.34)–(3.38), we obtain:

$$\begin{aligned}&E_k(l_1,l_2,...,l_n)=\delta(0,0,...,0)+\sum_{i=1}^{n}l_i\Delta\delta_i+A_k\left[H(0,0,...,0)+\sum_{i=1}^{n}l_i\Delta H_i\right],\\&\delta(0,0,...,0)=\delta_{Fe}+k_1c,\\&H(0,0,...,0)=H_{Fe}+k_2c,\end{aligned} \tag{3.40}$$

---

[1]The situation changes in the case of the ordered distribution of the atoms over the selected nodes of the crystal lattice. Calculation of anisotropic, i.e. depending on direction, contributions to the shift of the lines of the hyperfine splitting (magnetic and electrical quadrupole) requires examination of the specific variants of the arrangement of l_i atoms in the i-th coordination sphere of the resonant isotope. The solutions of these problems were obtained in [52, 53]).

where l_i is the number of the atoms of the impurity in the i-th coordination sphere of the iron atom.

The Mössbauer spectra of iron with the elements, situated to the left of iron in the periodic table of elements (Al, Si, Mn, V, etc, $c \sim 5 \div 10$ at%), are characterised by the form shown in Fig. 1.4 *b*. They indicate the presence of satellite lines with the value $H_{ef} < H_{Fe}$ which is determined by the strongest effect of the first coordination sphere (presence of $l_1 = 0, 1, 2...$ atoms of the impurity in the vicinity of iron). In contrast to these impurities, the atoms of Co and Ni increase the strength of the effective magnetic field.

The relationship (3.40) may be presented in the following form:

$$E_k(l_1, l_2, ..., l_n) = E_k(0,0,...,0) + \sum_{i=1}^{n} l_i \Delta E_{ki}, \tag{3.41}$$

where $E_k(0,0,...,0) = \delta(0,0,...,0) + A_k H(0,0,...,0)$ and $E_{ki} = \Delta\delta_i + A_k \Delta H_i$. The signs of $\Delta\delta_i$ and ΔH_i are determined by the nature of hyperfine interactions in the specific alloy.

In the absence of spontaneous magnetisation of the material and also external fields, only the term associated with the isomeric shift (Fig. 5.2) remains. The possibility of separating the individual sub-spectra and of determining their intensity indicates in fact the possibility of examining of the atomic and magnetic structure of the alloys using the Mössbauer effect. These investigations are described in Chapter 5.

In a disordered alloy, the probability of various configurations of the environment of the resonant atom by the atoms of the second component $P(l_1, l_2, ..., l_n)$ is determined by the relationship:

$$P(l_1, l_2, ..., l_n) = \prod_{i=1}^{n} P(l_i). \tag{3.42}$$

The probabilities $P(l_i)$ are given by the binomial distribution:

$$P(l_i) = \frac{z_i!}{(z_i - l_i)! l_i!} c^{l_i} (1-c)^{z_i - l_i}, \tag{3.43}$$

where c is the concentration of the impurity in the atomic fractions. For multicomponent diluted solid solutions identical relationships may be put down.

The equations (3.40) and (3.43) specify the position and intensity of the lines, i.e. they determine the form of Mössbauer spectrum. In the explicit form, the shape of the Mössbauer spectrum in the identical model is given in Chapter 4 (equation (4.40)).

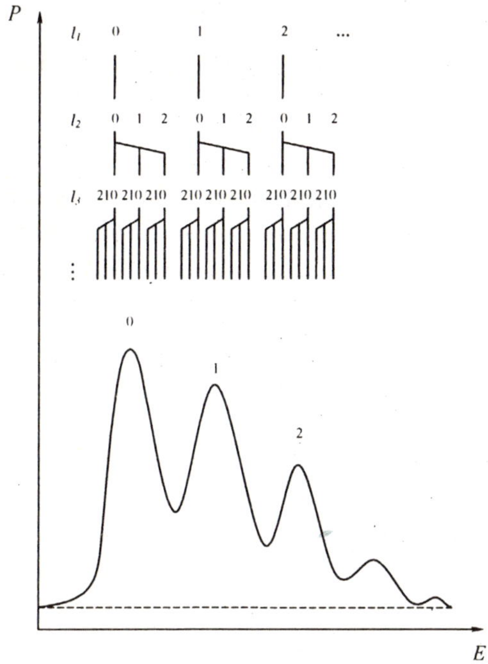

Fig. 3.5 Diagram illustrating the origin of quasi-continuous distribution $P(E)$ (the variation of resonant energy E depending on the number of impurity atoms in first (l_1), second (l_2) and third (l_3) coordination spheres of iron atoms).

The relationships (3.40), (3.41) indicate that the solid solutions are characterised by the quasi-continuous distribution of the energies of the spectrum components. Figure 3.5*a* shows the scheme illustrating the quasi-continuous distribution of energy for the components of the external peak $-1/2 \rightarrow -3/2$ in the case of dissolution of non-magnetic impurities of the type of Al, Si in iron. It should be mentioned that taking into account (3.36) the dependence of ΔE_{ki} on the number of the coordination sphere i is of the oscillating nature (see also paragraph 3.2 and Fig. 3.2).

The position of the centres of gravity of the satellite lines (Fig. 3.5), determined by the strongest effect of the first coordination sphere (more remote spheres are manifested in broadening and some shifting of these lines), may be determined by averaging using the relationships (3.40)–(3.42):

$$E_k(l_1) = \delta_{\text{Fe}} + A_k H_{\text{Fe}} + c\left[k_1 + A_k k_2 + \sum_{i=2}^{n} z_i \left(\Delta\delta_i + A_k \Delta H_i\right)\right] + \\ + l_1\left(\Delta\delta_1 + A_k \Delta H_1\right) \tag{3.44}$$

or $E_k(l_1)=E_k+cq_k+l_1(\Delta\delta_1+A_k\Delta H_1)$, where $l_1=0, 1, 2, 3...$

3.5. ANTIFERROMAGNETICS AND NON-COLLINEAR MAGNETICS

In the previous section, we examined the impurity crystals on the basis of ferromagnetic transition metals, being systems with mainly ferromagnetic interactions, although many impurities (see Table 3.2) interact in the antiferromagnetic manner with the atoms of the ferromagnetic matrix. In the metals and alloys in which all exchange interactions are negative, the antiferromagnetic ordering of the atomic magnetic moments takes place below the Neel temperature (T_N). Different types of magnetic ordering may occur in the presence of mixed exchange interactions (*i.e.* exchange interactions of different sign, for example, $J_{A-A}>0$, $J_{A-B}<0$, $J_{B-B}>0$ in a binary alloy). Ferromagnetic (FM), antiferromagnetic (AFM) and also non-collinear spin states, such as asperomagnetic (AS) and anti-asperomagnetic (AAS) may form, depending on the values of the exchange integrals, temperature, the composition and features of distribution of the atoms (section 5.3.2).

The main mechanisms of the formation of the magnetic field on the nuclei of the atoms remain the same as those examined on the example of crystals of transition metals with impurities. At the same time, in concentrated alloys it is necessary to take into account the effects of saturation of charge and spin density. The latter results in the non-additive nature of the contributions to the isomeric shift and the effective magnetic field on the nucleus. In addition to this, instead of the presence of only two possible directions of the moments, which is characteristic of collinear magnetics, it is necessary to take into account the arbitrary directions of the nodal magnetic moments. Regardless of the complicated nature of this problem, if all the effects are accurately taken into account, it is possible to select from the group of the permissible models (formed on the basis of the available a priori information) the optimum model adequately describing the Mössbauer spectrum, and identify the type of magnetic structure [54, 456].

Examples of the solution of similar problems in specific alloys are presented in Chapter 5.

Chapter 4

STRUCTURE OF MÖSSBAUER SPECTRA OF ALLOYS WITH LONG-RANGE AND SHORT-RANGE ATOMIC ORDER

The Mössbauer effect is suitable as an experimental method of determination of the quantitative characteristics of atomic distribution (the parameters of short-range and long-range orders) because of the high sensitivity of the nuclei of the resonant isotope to the local environment.

As shown in chapters 1–3, the non-equivalent positions of the resonant atoms in the crystal differ in the parameters of chemical interaction which, because of the interaction of the electron and nuclear subsystems, is manifested in the variation of the energy of nuclear transitions and the formation of completely or partially resolved components in the Mössbauer spectrum.

In paragraph 2.4, we presented the generalised form of the Mössbauer spectrum of a solid solution (equation (2.5)). For the determination of the atomic structure of the crystal by the method of nuclear gamma resonance spectroscopy it is necessary, firstly, to determine the quantitative relationship between the parameters of the Mössbauer spectra, on the one side, and of the characteristics of distribution of the atoms, on the other side, and secondly, propose a method of correct determination of the order parameters on the basis of experimental gamma resonance spectra.

4.1. ON SPECIFICS OF INFORMATION OBTAINED IN CONNECTION WITH THE LOCAL-NUCLEAR NATURE OF THE METHOD

When discussing the possibilities of investigating the atomic ordering by the method of nuclear gamma resonance, it is necessary to pay special attention to a number of important moments. Let us first consider the problems associated with the possibilities of examining the short-range order.

Evidently, at present, the suitability of the Mössbauer effect for examining the short-range order is not doubted because of its local-nuclear nature and high sensitivity of the energy of nuclear transition to the variation of the local atomic environment of the resonant nuclei. However, a number of authors [56, 57] have justifiably raised questions, associated with the principal difference of the method of nuclear gamma resonance in comparison with diffraction methods, used traditionally for investigation of the short-range order.

In fact, whilst the phenomenon of coherent scattering is a typical collective phenomenon, the information, obtained on the basis of the Mössbauer effect is the result of averaging over a statistical ensemble of the individual absorbing resonant nuclei, distributed in the crystal. Therefore, special attention is given to the problems, associated with the application of the Cowley–Warren parameters for describing the probabilities of atomic configurations and the possibility of taking into account many-particle effects [56].

The authors of [56, 57] examined the problem of incorrect application of the binomial distribution in the calculation of the probabilities of local environment of the resonant nuclei, expressed by means of the parameters of the Cowley–Warren pair correlation. At the same time, when using the accurate (in its sense) expansion of the probabilities of figures in terms of the correlation moments, proposed in [58], serious difficulties arise in the determination of the required number of the terms of expansion and the evaluation of the errors, introduced in breaking an infinite series.

With the accuracy to linear terms, the conventional binomial expansion [59, 60] and the expansion [56] coincide. The latter indicates that when examining the initial stages of ordering, both methods of description are equivalent. It should also be mentioned that when using the binomial distribution (paragraph 3.4, equation (3.43)), in which the mean concentration is replaced by some effective concentration $c^* = c_B\left[1 - \frac{\varepsilon_{AB}(\rho)}{c_A c_B}\right]$ (see section 4.2) numerically equal to the probability of detection of the atom of type *B* in the vicinity of the atom of type *A* (on the condition that the site is occupied by atom *A*) in an alloy with a short-range order, we carry out a kind of an approximation, taking into account the terms of a high order infinitesimal. This approximation is sufficiently accurate even in the case of a high degree of ordering of the alloy. The restriction of the expansion [58] by the linear terms in an identical case may result in larger error.

It is also necessary to examine the complicated problem of the relationship of a specific structure ('type') of the short-range order (according to the currently available classifications [61, 62], there are homogeneous, 'microdomain', local, etc., types of short-range order) with special features of the shape of the resonance absorption curves. Therefore, it appears rational to mention the result obtained by the author of this book [63] for the relationship of the parameter of pair correlation (or, with the accuracy to the multiplier $c(1-c)$ – the Cowley–Warren parameter) with the mean effective size of the antiphase domain (for more detail, see Appendix 3). The respective analytical relationship has the following form:

$$\varepsilon_{AB} = c_A^2 + c_B^2 - \langle t \rangle^{-1} \quad (\varepsilon_{AB} = -\varepsilon_{AA} = -\varepsilon_{BB}), \tag{4.1}$$

where ε_{AB} is the parameter of pair correlation; $<t>$ is the effective mean size of the domain equal to the ratio of the volume of the crystal to the general length (area) of the antiphase boundaries; c_A and c_B are the concentrations of the components. As shown in [63] (Appendix 3), parameter $<t>$ has a fully determined geometrical meaning: with the accuracy to the shape factor, it is the mean minimum size of the antiphase domain (diameter in the case of the 'spherical' and 'cylindrical' domains, the thickness in the case of flat domains, and so on). The relationship (4.1) is accurate and removes the problem of 'visualisation' of the pattern of atomic ordering because the equivalence of the 'probability' and 'topological' models is established. Thus, in the model of pair interactions, Mössbauer measurements make it possible to determine the parameters of pair correlation $\varepsilon_{AB}(\rho_i)$ or the Cowley parameter $\alpha_i = \frac{\varepsilon_{AB}(\rho_i)}{c_A c_B}$ $(-1 \le \alpha_i \le 1)$ equivalent to this parameter. At $\alpha_1 < 0$

there is a surplus of pairs *AB* (short-range ordering), and at $\alpha_1 > 0$ there is a shortage (short-range separation).

In [64], the model of pair interactions was used for the theoretical description of the short-range atomic order on the background of the long-range order.

When an alloy has regions greatly differing in the degree of the short-range ordering (and the size of the domains), the important characteristic of the structure of the short-range order may be the *dispersion* $D(t)$ of the effective size of the domain whose value should also influence the form of the Mössbauer spectrum. In a general case, the problem of recovery of the domain size distribution density function $f(t)$ may be considered. In the model which takes this distribution into account, the Mössbauer spectrum may be represented in the form of superposition of continuously distributed sub-spectra, corresponding to regions with different values of the parameter of pair correlation. This and a number of other problems evidently require further examination.

Ending the brief discussion of the possibilities of obtaining information on the state of short-range order of the alloys by the method of nuclear gamma resonance, it should be mentioned that the Mössbauer effect, regardless of a number of restrictions resulting from its nature, has also a number of advantages. For example, the method may be used efficiently for evaluating the values of the correlation parameters even in cases in which this problem cannot be solved using the methods such as x-ray and electron diffraction (for example, in the case of close in value atomic factors of scattering of components of the alloys, and also in examination of diluted alloys).[1] This also relates to the initial stages of atomic ordering in which the regions of coherent scattering are small (smaller than 8÷10 nm) and no sites of the reciprocal lattice form. In a general case, to obtain detailed information on the type and degree of the short-range order and its structural special features (the size and shape of domains), it is obviously necessary to use different experimental methods, supplementing each other.

Another relatively complicated problem is the problem of the 'information content' of the method of nuclear gamma resonance in respect of the long-range order taking into account the fact, as already mentioned, of the local–nuclear nature of the method and the sensitivity of the resonant nuclei only directly to the nearest environment.

---

[1]In many cases, the Mössbauer effect is also more efficient than the method of the neutron diffraction.

According to the current views regarding the long-range order, its formation indicates the preferential population of the sites of infinite separated sub-lattices by the atoms of a specific type[1]. The formation of the long-range order (for definiteness, it is assumed that the short-range order does not form on the background of the long-range order) is accompanied by changes in the fraction of the resonant nuclei in different sublattices and this is reflected unavoidably in the variation of the integral intensities of the components of the Mössbauer spectrum, corresponding to different configurations of the local environment of the Mössbauer nuclei.[2]

At the same time, whilst the regions of coherent scattering are small (smaller than 8÷10 nm), the superstructural lines do not form on diffraction patterns. In this sense, the Mössbauer effect has advantages in comparison with the diffraction methods. To avoid misunderstanding, it should be mentioned that in determination of the type and degree of the long-range order we proceeded from the existence of the relationship between the values of the order parameters and the parameters of the Mössbauer spectrum and did not link our considerations with the size of blocks of coherent scattering.

For the superstructures with stoichiometric composition and ideal ordering, the structure of the Mössbauer spectrum is determined by the presence of a small number s (usually s = 1, 2 or 3) of positions of the resonant nuclei, with the characteristic (for each of the superstructures) fully determined atomic environment in the nearest and subsequent coordination spheres. The probabilities of population of the sites of the sublattices (correspondingly the fractions of the nuclei in different positions) for possible types of superstructures may be easily calculated and compared with the experimental data. The presence, in a Mössbauer spectrum, of distinctively separated sub-spectra with the respective ratio of the integral intensities makes it possible to judge reliably about the presence of long-range order in the alloy.

As already mentioned, on the basis of the results of Mössbauer

[1]Evidently, this idealised model does not actually correspond to the real pattern of ordering of the alloy, but analysis of the initial considerations is outside the framework of this book.

[2]In a general case, the integral intensities of the sub-spectra may be expressed (see paragraph 4.3) through the probabilities of population of the sites of different sub-lattices by the atoms of the components of the alloy (or by means of the parameters of the long-range order connected with them by a linear relationship). This forms the analytical base of the applicability of the method of nuclear gamma resonance for investigation of the long-range order in the alloys.

measurements it is difficult to judge about the size of the blocks of coherent scattering (and the formation of nodes of the reciprocal lattice), *i.e.* the formation of the superstructures in the conventional sense. Consequently (in particular, in the initial stages of ordering) it is reasonable, as in the case of the short-range order, to use methods supplementing each other.

4.2. ON DETERMINATION OF THE ORDER PARAMETERS

It should be mentioned that it is quite simple to determine formally the fractions of the nuclei of the resonant isotope n_j in different non-equivalent positions (or the probability of non-equivalent positions P_j). As indicated by equation (2.5), for this purpose it is sufficient to decompose the Mössbauer spectrum into individual components and calculate the ratio of the integral intensities (areas) S_j. In a general case, it is necessary to take into account the difference in f_j and the effect of 'saturation of the area', *i.e.* the finite thickness of the absorber. However, this procedure is evidently not equivalent to the identification of the type and degree of the order. The point is that the probabilities P_j, characterising the distribution of the atoms in N nearest coordination spheres, are not independent and can be expressed through a considerably smaller number of the order parameters[1]. Of course, the aim of Mössbauer investigation of ordering alloys should not be the determination of the probabilities of different non-equivalent positions of the resonant isotope P_j, whose number in the general case is relatively large, but it should be the determination of the physical parameters of the order, describing the distribution of the atoms in the solid solutions.

It might seem that the problem of determination of the order parameters can be divided into two stages: 1. Determination of values of S_j and, consequently, 2. Determination of the order parameters, for example, using the method of maximum likelihood. However, this approach is in principle not correct because the method of maximum likelihood should be applied directly to the measured values [65]. In addition to the fact that such estimates are not substantiated, another disadvantage of this method is that it greatly complicates the problem.

The above considerations show that it is necessary to use a different approach, guaranteeing the reliable determination of the order parameters and other quantities directly from the Mössbauer

[1]The problem of the number of independent order parameters in the model of pair interactions has been examined in [64].

spectra; at that, their estimates should be optimum from the viewpoint of statistics. Since the method of maximum likelihood (see Appendix 1) requires that the law of distribution of the errors and the type of approximating functions are available, in the examined case it is necessary, in particular, to know the theoretical form of the Mössbauer spectrum as a function of the physical parameters, including order parameters. The knowledge of the theoretical form of the Mössbauer spectrum, as discussed in detail in Appendix 1, is a necessary prerequisite for quantitative analysis.

The effect of atomic ordering on the intensity of the components of the Mössbauer spectra was described in a number of studies [5, 59, 60, 66–71]. These studies contain original concepts and examine important problems, associated with the study of the atomic order by the method of nuclear gamma resonance. The present chapter is concerned with describing the data, relating to the effect of ordering on the integral intensities, width and shift of the components of the Mössbauer spectra, the mean value of the effective magnetic field, and also contains detailed examination of the problems, associated with the identification of intermediate types of ordering (superstructures of higher ranks whose formation cannot be revealed by the conventional methods).

4.3. VARIATION OF INTEGRAL INTENSITIES OF SPECTRA COMPONENTS IN THE FORMATION OF SHORT-RANGE AND LONG-RANGE ATOMIC ORDER IN SUBSTITUTIONAL SOLID SOLUTIONS

Equation (2.5) shows that the integral intensities of the components of the Mössbauer spectra are directly linked with the probabilities $P(j;\ x_2,x_3,\ldots,x_s) \equiv P\ (l_1,\ l_2,\ldots,l_n;\ x_2,\ x_3,\ldots,x_s)$ of the environment of the resonant isotope formed by impurity atoms (l_i is the number of the atoms of the impurity in the i-th coordination sphere of the resonant isotope). Let us consider the relationship of P_j with the parameters of the long-range and short-range atomic orders.

The long-range order

The symmetry of the crystal lattice and the limited extent of atomic interactions shows the possibility of the formation of only a finite number of different superstructures in crystals [72]. In the examination of the problem of the effect of the long-range order on the parameters of the Mössbauer spectrum of iron-based binary alloys, containing from 0 to 25 at% of the alloying element (substitutional impurity) [5, 67–71], we took into account the possibility of realisation of all ordered structures permitted for the given concentration range, *i.e.* superstructures of the ranks 1, 2, 3, described by the formula $Fe_{16-k}X_k$ (k = 1, 2, 3, 4 and 8).

In order to increase the accuracy of the initial model, we calculated the probabilities of substitution of the nodes by the Gorskii–Bragg–Williams model taking into account the interactions in the two nearest coordination spheres (the method gives the results which are in agreement with the experimental data in a wide temperature and concentration range [73]).[1] The minimum of the configuration part of the free energy:

$$\Phi = \Phi_0 + \sum_{l=1}^{4}\left[P_x^l \ln P_x^l + \left(1-P_x^l\right)\ln\left(1-P_x^l\right)\right] - \frac{1}{8}\left(P_x^1 + P_x^2 - P_x^3 - P_x^4\right)\times$$
$$\times\frac{8W_1 - 6W_2}{kT} - \frac{6W_2}{kT}\left[\frac{1}{16}\left(P_x^1 + P_x^2 - P_x^3 - P_x^4\right)^2 + \frac{1}{4}\left(P_x^1 + P_x^2\right)\right], \quad (4.2)$$

was determined using an M-222 computer. Here $\Phi = F/(N_0\ kT)$, $\Phi_0 = E_0/(N_0\ kT)$, (F is the free energy of the alloy, E_0 is the internal energy of the non-ordered solution), P^l_x is the probability of occupation of the sublattices by impurity atoms.

The results of calculations indicate that for the concentrations not exceeding 25 at%, $P^1_x = P^2_x \approx P^3_x$ (Fig. 4.1 and 4.2).[2] This makes it possible to restrict our considerations to the introduction of a single order parameter, taking into account the variation of the concentration of the second component only in 'intrinsic' nodes. The introduced notations of the sublattices are shown in Fig. 4.2.

The long-range order parameter is represented by the parameter α numerically equal to the fraction of the atoms of the second component X, situated in the nodes A (Fig. 4.2). The transition to the Gorskii–Bragg–Williams long-range order parameter is carried out using the equation $\eta = c(\alpha-n)/[n(1-n)]$, where c is the concentration of the second component in atomic fractions. In the accepted model, the distribution of the atoms in the nodes of the crystal lattice in the presence of a long-range order in the alloy is characterised by the probabilities:

[1]Calculations were carried out for the Fe–Si system, characterised by relatively high values of ordering energy (the ordering energy in the first and second coordination spheres is: $W_1 = 2.77\cdot10^{-20}$ J, $W_2 = 1.38\cdot10^{-20}$ J [73].

[2]This is valid for the majority of the ordering iron alloys. This pattern is determined by the values of W_1 and W_2.

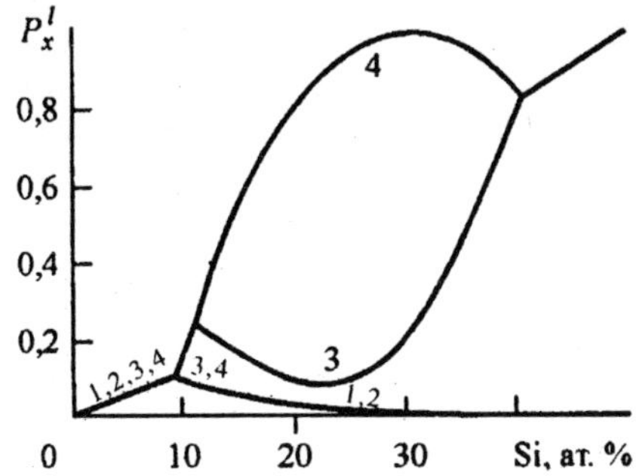

Fig. 4.1 Probabilities of occupation of sub-lattices by silicon atoms, calculated by the Gorskii-Bragg-Williams method (T = 543 K).

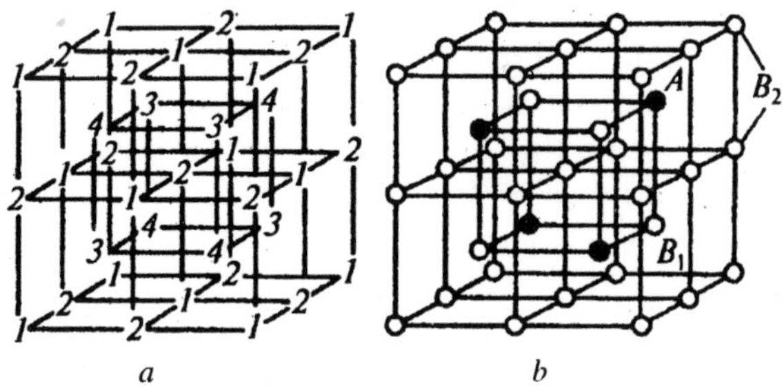

Fig. 4.2 Four FCC sub-lattices of the second rank (*a*) and notations used (*b*).

$$P(\text{Fe} \in B_1 \cup B_2) = \frac{1-n-c(1-\alpha)}{1-c}; \tag{4.3}$$

$$P(\text{Fe} \in A) = \frac{n-c\alpha}{1-c}; \tag{4.4}$$

$$P(\text{Fe} \in B_1 / \text{Fe} \in B_1 \cup B_2) = \frac{1/2-n}{1-n}; \tag{4.5}$$

$$P(\text{Fe} \in B_2 / \text{Fe} \in B_1 \cup B_2) = \frac{1/2}{1-n}; \tag{4.6}$$

$$P(B_1 \cup B_2 \ni \text{Si}) = \frac{c(1-\alpha)}{1-n}; \tag{4.7}$$

$$P(A \ni \text{Si}) = c\alpha / n. \tag{4.8}$$

Here c is the content of the impurity in the alloy (atomic fraction); $\ni, \in, \cup$ are the symbols denoting 'contains', 'belongs to' and 'union', respectively; α is the fraction of the atoms of the second component, located in the nodes A. The equations (4.5) and (4.6)

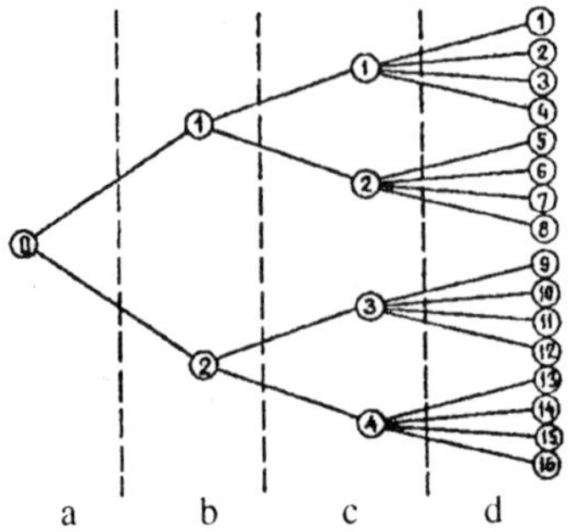

Fig. 4.3 Diagram showing the division of the initial BCC structure into sublattices: *a*) BCC structure; *b*) two simple cubic sublattices (superstructure of the first rank *AB*); *c*) four FCC sublattices with double spacing (superstructures of the second rank A_3B, AB and AB_3); *d*) sixteen simple cubic sublattices with the double spacing (superstructures of the third rank $A_{15}B$, A_7B, $A_{13}B_3$, $A_{11}B_5$, A_5B_3, A_9B_7, ..., AB_{15}) [16].

represent the conditional probabilities, calculated assuming that the distribution of iron in the sublattice $B_1 \cup B_2$ is statistically uniform.

Using the equations (4.3)–(4.8), the methods of combinatorics may be used to calculate the probability of presence of l_i atoms of the impurity in the i-th coordination sphere of iron. The separation of the probabilities in the division of the crystal lattice of the ordering alloy to sublattices (Fig. 4.3) generates summation in over different sublattices. For $i = 1$, the expression describing the dependence of the probability of the resonant isotope surrounding by l_1 atoms atoms of the second component on the concentration and the parameters of the long-range order has the following form [68]:

$$P^k(l_1,\alpha) = \frac{1-n-c(1-\alpha)}{1-c}\left[\frac{1/2-n}{1-n}C_{z_1}^{l_1}\left(\frac{c(1-\alpha)}{1-n}\right)^{l_1}\left(1-\frac{c(1-\alpha)}{1-n}\right)^{z_1-l_1}+\right.$$

$$+\frac{1/2}{1-n}\left.\begin{bmatrix} \sum\limits_{j=0}^{l_1} P(j)P'(l_1-j), & 0 \le l_1 < k+1, \\ \sum\limits_{j=0}^{k} P(j)P'(l_1-j), & k+1 \le l_1 \le z_1-k, \\ \sum\limits_{j=l_1-z_1+k}^{k} P(j)P'(l_1-j), & z_1-k < l_1 \le z_1 \end{bmatrix}\right] + \frac{n-c\alpha}{1-c}\times \qquad (4.9)$$

$$\times C_{z_1}^{l_1}\left(\frac{c(1-\alpha)}{1-n}\right)^{l_1}\left(1-\frac{c(1-\alpha)}{1-n}\right)^{z_1-l_1}, \quad \alpha \ge n,$$

where

$$P(j)=C_k^j\left(\frac{c\alpha}{n}\right)^j\left(1-\frac{c\alpha}{n}\right)^{k-j};$$

$$P'(l_1-j)=C_{z_1-k}^{l_1-j}\left(\frac{c(1-\alpha)}{1-n}\right)^{l_1-j}\left(1-\frac{c(1-\alpha)}{1-n}\right)^{z_1-k-l_1+j},$$

where j is the number of atoms X, situated in the nodes A in the first coordination sphere of iron (iron is in node B_2), z_1 is the coordination number for the first coordination sphere. Equation (4.9) satisfies the normalisation condition and at $\alpha = n$ it transforms to the binomial law of distribution of the probabilities, characterising the non-ordered state.

The otained equation at constant f_q' gives the dependence of the intensities of the components of the Mössbauer spectrum (with the accuracy to the multiplier, transferring the probability to the number of counts in the channel of the spectrometer) on the concentration and degree of long-range order in the alloy in the model of a single coordination sphere. To determine the integral intensities, *i.e.* the areas of the partial components of the spectrum S_q, it is important to take into account the variation of the width of the lines of these components[1]. When taking into account the differences in f_q' evidently, $S_q \sim f_q' P_q$. In a general case, it is also necessary to take into account the finite thickness of the absorber (the effect of saturation of the area) by the introduction of the appropriate corrections which may be calculated by the method proposed in [74].

The values of the probabilities $P^k(l_i,\eta)$ for i = 2, 3,... may be obtained similarly using the representation of trees or graphs of possible outcomes [75] and contain, like equation (4.9), summation over different sublattices.

The relationship of the integral intensities of the components of the Mössbauer spectra with the parameters k and η, describing the type and degree of the long-range order, is, as can be seen, relatively complicated. This is caused by the fact that in the Mössbauer measurements the information on the distribution of the atoms is obtained from the resonant nucleus and from its nearest environment. The description of the intensity of the lines of the spectrum in the presence of long-range order is equivalent to establishing the relationship between the characteristics of the local and long-range atomic order. This requires introduction of chains

[1]The effect of ordering on the width of the lines is taken into account in paragraph 4.4.

of conditional probabilities and calculation of lattice sums.

This examination takes into account the possibility of the formation in the BCC iron-based alloys with the concentration of the second component ≤25 at.%, of all principally possible superstructures of the first, second and third rank ($Fe_{16-k}X_k$; k = 1, 2, 3, 4 and 8). Examination results can be generalised for the alloys with a higher content of impurities (see investigation of the iron alloys with 37.5 and 40 at.% of aluminium, paragraph 5.1, Figs. 5.2 and 5.3).

The short-range order

The formation of the short-range order also results in the variation of the probabilities of environment of the resonant isotope by the components of the alloy and is manifested in the change of the integral intensities of the Mössbauer spectra S_j. Let us consider the relationship of S_j with the parameters of the short-range order in the model which takes into account only the pair interactions in the crystal.

Let y be the atoms of the Mössbauer isotope, and x be the atoms of the second component. In the Mössbauer measurements, the information is obtained from the sites, occupied by the atoms y and, consequently, it is necessary to use the conditional probability:

$$P_x(\rho_i) = \frac{P_{xy}(\rho_i)}{c_y}, \tag{4.10}$$

where $P_{xy}(\rho_i)$ is the probability of the presence of the pair xy at distance ρ_i, which may be expressed by the parameter of the pair correlation $\varepsilon_{xy}(\rho_i)$ and the concentration of the components. According to the definition [76]:

$$\varepsilon_{xy}(\rho_i) = P_{xy}(\rho_i) - c_x c_y. \tag{4.11}$$

The previously introduced probability $P_x(\rho_i)$ is in fact the effective concentration of the impurity in the i-th coordination sphere of the resonant nucleus.

Thus, from the Mössbauer measurements we can determine the effective concentration $P_x(\rho_i) = c_x\left[1 + \frac{\varepsilon_{xy}(\rho_i)}{c_x c_y}\right]$, which is linked with the parameter of the pair correlation $\varepsilon_{xy}(\rho_i)$ and the probability of the presence of l_i impurity atoms in the i-th coordination sphere of the resonant atom:

$$P\left(l_i,\varepsilon_{xy}(\rho_i)\right)=C_{z_i}^{l_i}\left\{c_x\left[1+\frac{\varepsilon_{xy}(\rho_i)}{c_xc_y}\right]\right\}^{l_i}\left\{1-c_x\left[1+\frac{\varepsilon_{xy}(\rho_i)}{c_xc_y}\right]\right\}^{z_i-l_i}. \qquad (4.12)$$

A similar description for the case of the short-range order was used in [59, 60, 25]. Thus, the Mössbauer measurements may be used to determine the parameter of pair correlation $\varepsilon_{xy}(\rho_i)$ of the equivalent Cowley (or Cowley–Warren) parameter $\alpha_i=-\frac{\varepsilon_{xy}(\rho_i)}{c_xc_y}$ $(-1\le\alpha_i\le1)$. At $\alpha_1<0$, there is an excess of the pairs xy (short-range ordering), and at $\alpha_1>0$ there is a shortage (short-range separation). The probability of the configuration $l_1, l_2,\ldots, l_N$ in the presence of the short-range order in the alloy can be written in the form of the product:

$$S_j\sim P(l_1,l_2,\ldots,l_N;\alpha_1,\alpha_2,\ldots,\alpha_N)=\prod_{i=1}^{N}P(l_1,\alpha_i). \qquad (4.13)$$

The relationships (4.12), (4.13) and (2.5) make it possible to model the form of the γ-resonance spectrum in the presence of the short-range order in the alloys.

Analysis of the models of pair interactions may be used as the starting point for different refinements. For example, in the study by Sidorenko *et al* [58,77], the authors proposed methods for determination of many-particle atomic correlations on the basis of analysis of the experimental gamma resonance spectrum. In these studies, the authors obtained for the first time the experimental values of the correlation moments of the third order ($Fe_{1-x}Si_2$ and $Fe_{1-x}Al_x$ systems were investigated).

4.4. ACCOUNTING FOR THE EFFECT OF REMOTE COORDINATION SPHERES IN THE PRESENCE OF ATOMIC ORDER

We have presented the basic equations describing the probability of different non-equivalent positions of the resonant isotope in solid solutions in the realisation of the short-range and long-range order. Since the effects determining the variation of the energy of nuclear levels (paragraph 2.2) rapidly decrease with an increase of the distance of the impurity from the Mössbauer nucleus, in all actual cases it is rational to take into account the effect of only several nearest coordination spheres.

In order to increase the accuracy of determination of the parameters of the short-range and long-range order from the Mössbauer measurements, it is necessary to evaluate the effect of impurities of the remote coordination spheres on the width and displacement of the Mössbauer line. These attempts were made in [78] for an ideally non-ordered alloy. We examine a general problem of the displacement and widening of the absorption line in the presence of the short-range and long-range atomic order in an alloy.

We take into account the effect of the $(N+1)$-th coordination sphere. The effect of the atoms of the impurity of the $(N+1)$-th sphere leads to the dispersion of the values of the effective magnetic field and isomeric shift. For the resultant dispersion of the position of the l-th line of the Mössbauer spectrum (l = 1–6), the following relationship is valid:

$$D(x_l) = D(A_l H + \delta), \tag{4.14}$$

where

$$A_{1,6} = \pm\frac{1}{2}(\mu_e - \mu_g)(\mu_e - \mu_g)^{-1} = \pm\frac{1}{2};\quad A_{2,5} = \pm\frac{1}{2}\left(\frac{1}{3}\mu_e - \mu_g\right)(\mu_e - \mu_g)^{-1};$$

$A_{3,4} = \pm\frac{1}{2}\left(-\frac{1}{3}\mu_e - \mu_g\right)(\mu_e - \mu_g)^{-1}$; μ_e, μ_g are the magnetic moments of the nucleus Fe^{57} in the ground and excited states. The values of H, δ and of the position of the Mössbauer line x_l are measured in energy units.

For an absorber with a small effective thickness ($C_a = f'n\sigma_0 \ll 1$) the form of the curve of resonance absorption [6] in the absence of the effects of splitting and displacement of the nuclear levels may be described by the Lorentzian with the width $2\Gamma_0$. The split spectra represent the superposition of the Lorentzians.

While retaining the Lorentzian shape of the approximating line, the accuracy of determination of the physical parameters may be increased if we take into account the effect of configuration broadening. For this purpose, the central moments of the second order for the approximating line (1.5) and the line broadened as a result of the effect of the $(N+1)$-th sphere must be equal:

$$M_2' = M_2 + D(A_l H + \delta). \tag{4.15}$$

The main problem is that the Lorentzian

$$\varepsilon(x) = \frac{\varepsilon(x_0)}{1 + 4/\Gamma^2 (x - x_0)^2} \tag{4.16}$$

does not have the finite moment of the second order but since the expansion

$$e^{(x-x_0)^2/2\sigma^2} = 1 + 0 + \frac{1}{\sigma^2}\frac{(x-x_0)^2}{2} + \ldots, \tag{4.17}$$

is valid, we can use the approximate equality:

$$\frac{\varepsilon(x_0)}{1 + 4/\Gamma^2 (x - x_0)^2} = \frac{\varepsilon(x_0)}{1 + 1/(2\sigma^2)(x - x_0)^2} \approx \varepsilon(x_0) e^{-(x-x_0)^2/2\sigma^2}. \tag{4.18}$$

Taking into account that

$$M_2 = \int_{-\infty}^{\infty} (x - x_0)^2 e^{-(x-x_0)^2/2\sigma^2} dx = \sigma^2, \tag{4.19}$$

and also relationship (4.18), we obtain:

$$M_2 = \sigma^2 + D(A_l H + \delta) = \int_{-\infty}^{\infty} (x - x_0)^2 \exp\left[-\frac{(x - x_0)^2}{2\sigma'^2}\right] dx = \sigma'^2. \tag{4.20}$$

The equations (4.18) and (4.20) for the broadening of the lines finally give:

$$\Delta\Gamma_l \approx \frac{4}{\Gamma_l} D(A_l H + \delta), \tag{4.21}$$

and the problem of taking into account broadening is consequently reduced to the calculation of the dispersion of the position of the Mössbauer line $D(A_l H+\delta)$, sensitive to the variation of the order parameters.

The long-range order

From the definition of the dispersion as the central moment of the second order relative to the averaged-out model of N coordination spheres we have

$$D(A_l H+\delta)=\sum_{l_{N+1}=0}^{z_{N+1}} P^k(l_{N+1},\eta)\big[A_l H(l_1,l_2,...,l_{N+1})+\delta(l_1,l_2,...,l_{N+1})-$$
$$-A_l H(l_1,l_2,...,l_N,\overline{l}_{N+1})-\delta(l_1,l_2,...,l_N,\overline{l}_{N+1})\big]^2. \tag{4.22}$$

Using the relationships (3.37) and (3.38), the expression is easily transformed to the following form:

$$D(A_l H+\delta)=\sum_{l_{N+1}=0}^{z_{N+1}} P^k(l_{N+1},\eta)l_{N+1}^2-R_l^2Q^2, \tag{4.23}$$

where

$$R_l=A_l\Delta H_{N+1}+\delta_{N+1} \tag{4.24}$$

and

$$Q=\sum_{l_{N+1}=0}^{z_{N+1}} P^k(l_{N+1},\eta)l_{N+1}. \tag{4.25}$$

Carrying out combinatorial transformations using the generating functions of the moments $M(t)=\sum_r e^{rt}P_r$, we obtain the following results:

$$Q=z_{N+1}\left[1+\frac{n^2\eta^2}{c(1-c)}\right]c \tag{4.26}$$

and

$$D(A_l H+\delta)=R_l^2\left\{\left\{z_{N+1}c(1-c)+\frac{z_{N+1}n\eta^2}{(1-c)}\big[(c-\eta n-1)(1-2n)+\right.\right.$$
$$\left.\left.+n\big(z_{N+1}-c(z_{N+1}+1)\big)\big]-z_{N+1}^2\frac{n^4\eta^4}{(1-c)^2}\right\}. \tag{4.27}$$

Equation (4.21) can now be used to calculate the broadening of the Mössbauer line $\Delta\Gamma_l$, determined by the effect of the $(N+1)$-th sphere. Product QR_l is nothing else but the displacement of the Mössbauer line Δx_l (see 4.14).

For the non-ordered state ($\eta = 0$), calculations using equation (4.21) coincide with the results obtained in [78].

Analysis of the equations shows that the displacement and broadening of the Mössbauer lines, determined by the effect of the second and third coordination spheres, are relatively large (their value is governed by the values of ΔH_2, δ_2 and ΔH_3, δ_3). The correction for the broadening in the iron-based solid solutions may reach 50% of the initial width of the l-th line of pure iron Γ_l. The displacement for the first and sixth peaks ($l = 1, 6$) is several percent of the value of the effective magnetic field in pure iron.

The short-range order

In the case of the short-range order, the initial expression for the calculation of the dispersions differs from the relationship (4.22) by the fact that averaging is carried out in respect of the set of probabilities $P(l_{N+1}, \alpha_{N+1})$, determined by equation (4.12). The procedure used for averaging in this case is considerably simpler and gives the following relationship for the broadening of the Mössbauer lines:

$$\Delta\Gamma_l \approx \frac{4}{\Gamma_l} R_l^2 c(1-\alpha_{N+1})\left[1-c(1-\alpha_{N+1})\right] z_{N+1}. \tag{4.28}$$

At $\alpha_{N+1} = 0$, the equation coincides with the calculations carried out in [78]. The shift of the Mössbauer line is given by the expression:

$$\Delta x_l = QR_l = z_{N+1} c(1-\alpha_{N+1}) R_l. \tag{4.29}$$

As in the case of the long-range order, the values of $\Delta\Gamma_l$ and Δx_l are relatively high and this must be taken into account in the investigations.

4.5. RELATIONSHIP OF THE MEAN VALUE OF THE EFFECTIVE MAGNETIC FIELD WITH THE PARAMETERS OF THE LONG-RANGE AND SHORT-RANGE ATOMIC ORDER

As it follows from the general considerations, the value $\langle H \rangle$ should depend on the parameters of the short-range and long-range order[1] and, consequently, may be used as a characteristic carrying information on the distribution of the atoms in the alloys. Let us establish the relationship of <H> with the order parameters.

The long-range order

In the model of N coordination spheres the expression for the mean value of the magnetic field on the nuclei of ^{57}Fe atoms in the presence of a long-range order in the alloy (see (3.37) and (4.9)) may be written in the following form:

$$\langle H\rangle = H_0 + \sum_{i=1}^{N} \Delta H_i \sum_{l_i=0}^{z_i} l_i P^k(l_i,\eta). \tag{4.30}$$

As shown in the previous paragraph (equation (4.26)) the combinatorial transformations result in the following:

$$\sum_{l_i=0}^{z_i} l_i P^k(l_i,\eta) = z_i c\left[1+\frac{n^2\eta^2}{c(1-c)}\right], \tag{4.31}$$

where $c' = c\left[1+\frac{n^2\eta^2}{c(1-c)}\right]$ has the meaning of some effective concentration of the impurity in the alloys; for the non-ordered state $c' = c$. Equation (4.30) may be consequently reduced to the simple form:

$$\langle h\rangle = h_0 + c\left[1+\frac{n^2\eta^2}{c(1-c)}\right]\sum_{i=1}^{N} h_i z_i, \tag{4.32}$$

where $\langle h\rangle = \frac{\langle H\rangle}{H_{\text{Fe}}}$, $h_0 = \frac{H_0}{H_{\text{Fe}}}$, $h_i = \frac{\Delta H_i}{H_{\text{Fe}}}$ (H_{Fe} is the strength of the field in pure iron). The resultant relationship gives the dependence of the mean field value on the concentration and the long-range order parameter for different possible types of ordering. The type (rank) of the superstructure is determined by the parameter n. At $S = \sum_{i=1}^{N} h_i z_i < 0$, the value <h> decreases with

[1]At low impurity concentrations, the contribution of configuration effects is dominant.

increase of η, and at $S > 0$ it increases. The observed changes $\delta\langle H\rangle = \delta\langle h\rangle\ H_{Fe}$ for the iron-based alloys may reach 10÷15 kOe, which is an order of magnitude higher than the absolute error in the determination of $\langle H\rangle$. In relative measurements, the error of the experimentally determined relative changes $\delta\langle H\rangle$ may be greatly reduced.

Equation (4.32) shows that at the order ↔ disorder transition temperature, the curve of the dependence of the mean field on annealing temperature should have a knee if there is a phase transition of the second order, and a step at phase transition of the first order. This may be used for the identification of the type of order ↔ disorder phase transition.

The short-range order

In the presence of a short-range order in the solid solution, the mean value of the effective magnetic field is given by the following relationship:

$$\langle h\rangle = h_0 + \sum_{i=1}^{N} h_i \sum_{l_i=0}^{z_i} l_i\, p(l_i, \alpha_i), \tag{4.33}$$

which, taking into account expression (4.12) is reduced to the following form:

$$\langle h\rangle = h_0 + c\sum_{i=1}^{N} z_i h_i (1-\alpha_i). \tag{4.34}$$

Since

$$\frac{1}{h_k}\frac{\partial\langle H\rangle}{\partial\alpha_k} = -H_{Fe} z_k c < 0, \tag{4.35}$$

the sign of the derivative is determined only by the sign of h_k, and the sign of the addition to the mean field from the atoms of the impurity of the k-th coordination sphere in the presence of the short-range order in the alloy is opposite to the sign of the product $\alpha_k h_k$. Consequently, it is possible to evaluate the variation of the the mean field value in the short-range ordering and separation.

In the case of the short-range order with the sign of $S = \sum_{i=1}^{N} h_i z_i$ being constant, the value $\delta\langle H\rangle$ may assume both positive and negative values, depending on whether there is a short-range separation or ordering. For the absolute value of $\delta\langle H\rangle$ the estimates obtained for the long-range order are valid.

4.6. ON EXPERIMENTAL DETERMINATION OF THE VALUE OF THE MEAN FIELD

The value of $\langle h\rangle$ may be determined experimentally by the position of the centres of gravity of the external peaks ($-1/2 \to -3/2$; $+1/2 \to +3/2$) of the Mössbauer spectrum. The position of the centre of gravity of the peak may be determined most accurately using the expression:

$$\langle v\rangle = \sum_{j=N_1}^{N_2} v_j \varepsilon(v_j) \left(\sum_{j=N_1}^{N_2} \varepsilon(v_j) \right)^{-1}, \tag{4.36}$$

where N_1 and N_2 are the numbers of the channels of the beginning and end of the peak; $\varepsilon(v_j)$ is the effect, corresponding to the velocity v_j. If the velocity is measured in mm/s, then:

$$\langle h\rangle = 0.094\left(\langle v\rangle_{+1/2\to+3/2} - \langle v\rangle_{-1/2\to-3/2}\right)[\text{rel.units}]. \tag{4.37}$$

The accuracy of calculations using equations (4.36) and (4.37) is especially high at a low content of the impurity in the alloy when the superposition of the 'tails' of internal lines of the spectrum is minimum, i.e. exactly in the concentration range in which the conventional methods of examining the atomic order are ineffective and the measurements of $\langle h\rangle$ may prove to be highly useful. At a higher concentration of the impurity, because of the increase of the degree of overlapping of the lines of the hyperfine structure, it is necessary to take into account the superposition of the 'tails' of the second and fifth lines.

Figure 4.4 shows theoretical dependences calculated for different values of η using equation (4.32) and also the values of $\langle h\rangle$ obtained from the experimental Mössbauer spectra using (4.37) [79]. The sum, included in (4.32), was calculated using the data published in [80] in which the effect of six nearest coordination spheres was taken into account.

Curve 1 describes the theoretically predicted change of the mean field for the case $\eta = 0$. The experimental points at $c > 10$ at.% are located below this curve, as expected, because according to the available data, the alloys containing more than 10 at.% Si are highly ordered. Curve 3, calculated for the case of complete

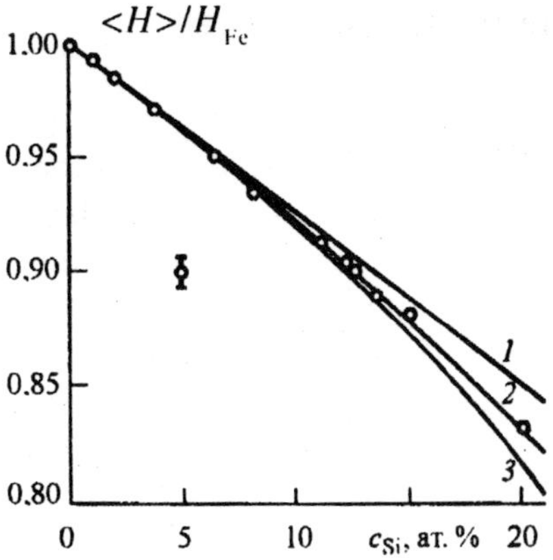

Fig. 4.4. Variation of the magnitude of the mean field with Si concentration. Points are experimental values. 1 – calculations for non-ordered alloys ($\eta = 0$, $\alpha = n$); 2 – calculations for an intermediate degree of the order ($\alpha = 0.6$); 3 – ($\alpha = 1$).

ordering ($\alpha = 1$), passes below the experimental points. The most accurate description is obtained at $\alpha = 0.6$ (see (4.9); the curves for different n are almost completely identical).

The presented data are an experimental confirmation of the sensitivity of $\langle h\rangle$ to the distribution of the atoms, although $\langle h\rangle$ is only slightly sensitive to the type of order (a weak dependence on n). The measurements of $\langle h\rangle$ may be used, for example, in plotting phase diagrams, combined with the complete processing of the reference spectra, and also when investigating the kinetics of ordering processes.

4.7 ANALYTICAL DESCRIPTION OF THE FORM OF MÖSSBAUER SPECTRA OF ORDERED ALLOYS

For diluted iron-based alloys, the short-range order parameters are best determined by the external peaks of the Mössbauer spectrum because they are best resolved and intensive. In concentrated alloys when overlapping of the lines of the Zeeman sextet is great, it is necessary to analyse the entire Mössbauer spectrum. In addition to this, at a high concentration of the impurity it is necessary to take into account the non-additivity of the contributions of the dissolved atoms in $H_{ef.}$

As shown in Chapter 3, at impurity concentrations up to 10÷15 at.% the additivity of contributions to the effective magnetic field and the isomeric shift from the atoms of the impurity closest to the iron is fulfilled with a high degree of accuracy. Then, in the model which takes into account the effect of the impurities of the first coordination sphere, the theoretical form of the first and sixth peaks of the Mössbauer spectrum in the presence of a long-range atomic order in the alloy may be written in the following form:

$$I(v_j)=I(\infty)\left[1-g^{(l)}\sum_{l_1=0}^{z_1}\frac{P^k(l_1,\eta)}{1+4/\Gamma^2\left(v_j-v_0^{(l)}-l_1\Delta v^{(l)}\right)^2}\right], \tag{4.38}$$

where l is the number of the peak (l = 1, 6); $I(\infty)$ is the level of the background; $g^{(l)}$ is the normalisation coefficient; Γ is the width of the line; $v_0^{(l)}$ is the position of the component corresponding to the iron atoms which do not contain impurity atoms in their nearest environment (this velocity should be measured in channels). The probabilities $P^k(l_1,\eta)$ are determined by (4.9). Parameters $\Delta v^{(l)}$ are linked with h_1 and δ_1 by the following equations:

$$\Delta v^{(1)}=-\frac{h_1H_{Fe}}{2\beta_1}+\frac{\delta_1}{\beta_2},\quad \Delta v^{(6)}=\frac{h_1H_{Fe}}{2\beta_1}+\frac{\delta_1}{\beta_2},$$

and

$$h_1=\left(\Delta v^{(6)}-\Delta v^{(1)}\right)\beta_1/H_{Fe};\ \delta_1=\left(\Delta v^{(6)}+\Delta v^{(1)}\right)\beta_2/2.$$

If H_{Fe} and δ_1 (magnetic field and isomeric shift) are measured in kOe and mm/s, respectively, β_1 transfers channels into kOe, and β_2 into mm/s. The values of the parameters $I(\infty)$, $g^{(l)}$, Γ, $\Delta v^{(l)}$, $v_0^{(l)}$, k and η for every peak are determined by finding the maximum of the non-linear functional:

$$\left\langle\varepsilon^2\right\rangle=\frac{1}{M-M_0+1}\sum_{j=M_0}^{M}\frac{1}{I^*(v_j)}\left[I(v_j)-I^*(v_j)\right]^2. \tag{4.39}$$

Here $I^*(v_j)$ is the experimental intensity of the Mössbauer spectrum, corresponding to velocity v_j; M_0 and M are the numbers of the channels of respectively the beginning and end of the peak. It should be mentioned that, generally speaking, $g^{(1)} \neq g^{(6)}$ because of the geometrical factor.

In the case of a short-range order, the form of the external peak is given by the relationship similar to (4.38), with the only difference being that the probabilities $P(l_1, \alpha_1)$ of the presence of l_i atoms of the second component in the first coordination sphere of the Fe atom are given by equation (4.12). When taking into account several coordination spheres, equation (4.13) is used. The procedure for determination of the parameters, including the Cawley parameters of the short-range order α_i, is the same as in the case of the long-range order.

For alloys with a higher concentration of the second component it is necessary to process the entire Mössbauer spectrum, as carried out in, for example, [59], where the formation of a short-range atomic order in the Fe + 13 at.% Cr alloy during its ageing at 450 °C was investigated.

Thus, when taking into account the effect of the two nearest coordination spheres in the case of a higher concentration of the second component (when the overlapping of the sextet lines is large), we can use the following relationships [59]:

$$I(v_j)=I(\infty)\left\{1-\sum_{i=1}^{6}\sum_{l_1=0}^{z_1}\sum_{l_2=0}^{z_2}a_i\times\right.$$
$$\left.\times\frac{P(l_1,l_2,\alpha_1,\alpha_2)}{1+4/\Gamma_i^2\left[v_j-\delta_0-l_1\delta_1-l_2\delta_2-A_iH_{\mathrm{Fe}}(1+K\,c)(1-l_1h_1-l_2h_2)\right]^2}\right\} \tag{4.40}$$

$$P(l_1,l_2,\alpha_1,\alpha_2)=C_{z_1}^{l_1}C_{z_2}^{l_2}\left[c(1-\alpha_1)\right]^{l_1}\left[c(1-\alpha_2)\right]^{l_2}\times$$
$$\times\left[1-c(1-\alpha_1)\right]^{z_1-l_1}\left[1-c(1-\alpha_2)\right]^{z_2-l_2}. \tag{4.41}$$

where i is the number of peak (i = 1–6); Γ_i is the width of the corresponding peaks of the Zeeman sextets of the lines; a_i are the intensity of the peaks (it is assumed that $a_i = a_{7-i}$); δ_0 is the isomeric shift for iron atoms whose first two coordination spheres contain no chromium atoms; δ_1 and δ_2 are the contributions to the isomeric shift from respectively the nearest and next nearest chromium atoms ^{57}Fe nucleus; A_i is the vector giving the position of the lines: –1/2 → –3/2, –1/2 → –1/2, –1/2 → +1/2, +1/2 → –1/2, +1/2 → +1/2, +1/2 → +3/2 (it is determined by the values of the moments of the ground μ_g and excited μ_e states of the ^{57}Fe nucleus); z_1 = 8 and z_2 = 6 are the coordination numbers of the first and second coordination spheres in the BCC lattice.

To determine the parameters $I(\infty)$, a_i, Γ_i, δ_0, δ_1, δ_2, K, h_1, h_2, α_1 and α_2 the minimum of the functional (4.39) was found by the standard *pattern search* method.

4.8 STRUCTURE OF MÖSSBAUER SPECTRA OF MULTILAYER SUPERLATTICES OF TYPE $\{A_m/B_n\}_k$

In the last years, in addition to alloys in which the short-range and long-range atomic order forms in a natural manner, special attention is given to study multilayer nanosized structures or superlattices. In some cases these are referred to as man-made superlattices. They form by consecutive 'deposition' of thin layers of different components (usually two – A and B). The thickness of these layers may be only several angströms. Recently, a large number of papers have been published on investigation of the structure and properties of such superlattices using the Mössbauer effect.

In contrast to solid solutions with short-range and long-range atomic order, the multilayer lattices have not as yet been described in detail as regards their atomic and magnetic structures. In particular, until recently, only a few studies have been carried out to examine the characteristics and population of different non-equivalent positions of resonant nuclei which greatly differ from those in solid solutions.

Some characteristics of the local and magnetic structures of the superlattices $\{A_m/B_n\}_k$ were determined by experiments and simulated in [81, 82, 330] (the results of experimental investigations are presented in Chapter 5).

In [82], the Monte Carlo method was used to simulate the population of single layers of the family of (100) superlattices $\{A_m/B_n\}_k$ with BCC symmetry. A populated fragment of the superlattice (Fig. 4.5) with the size $N \times N$ ($N \leq 20$) was assumed to be cyclically closed on itself in two dimensions. The coordinates $X, Y \in [1, N + 1)$ of atoms (sort A or B) 'falling' alternately on the oriented substrate were specified using a generator of random numbers. The incident atom occupied one of the nearest local energy minima, corresponding to the site of the lattice in the coordinates l, i, j (l is the number of the layer, $i, j = 1, 2, \ldots, N$), in one of the incompletely populated surface single layers of the growing film. The migration of the atom from the point of incidence to the populated site took place in the direction of the potential gradient. No unfilled sites (vacancies) existed because of this filling.

m layers of component B (X), n layers of the component A (^{57}Fe), were deposited on the substrate and then again m layers of component B, and so on. The numbers m and n could be artibrary integers or fractionall numbers, or they could be selected in such a manner that the shortage (or excess) of the atoms A and B at all

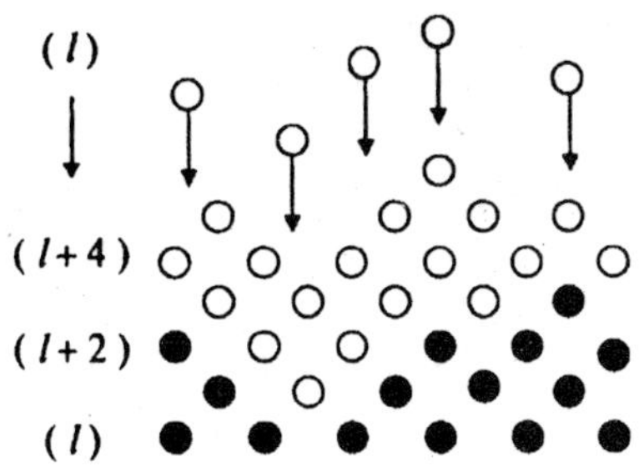

Fig. 4.5 Diagram of population of latttice sites using the Monte Carlo method (● – component A, ○ – component B).

interlayer boundaries remained constant. The processes of diffusion mixing of the components were not taken into account by the model (in practice the minimisation of the role of diffusion processes is achieved by reducing the temperature of the substrate to aproximately 150 °C). The magnetic interaction of the deposited atoms was not considered.

Therefore, calculations were carried out to determine the probability of configurations of the local environment of the atoms of type A by the atoms of type B which are of interest for interpreting the results of study of the interlayer boundaries in superlattices $\{A_m/B_n\}_k$ (in particular, in superlattices ^{57}Fe/Cr [330]) using the Mössbauer effect.

It was established that the distribution of the probabilities of the surrounding of the type A atoms (identical with resonant atom ^{57}Fe) by different numbers of type B atoms for superlattices of the type $\{A_m/B_n\}_k$ is of the 'saddle' character (Fig. 4.6*b*) (it is well known that the non-ordered solid solutions are characterised by the 'bell-shaped' distribution of the probabilities described by a binominal law, Fig. 4.6*a*). The change in the type of distribution from the 'saddle shape' to the 'bell shape' may take place as a result of the diffusion mixing of the components in the vicinity of interlayer boundaries.

Calculations were carried out to determine the matrices of probabilities $P\ (l_1,\ l_2)$ of surrounding the atoms A by different numbers of atoms B in the first and second coordination spheres for different values of the parameters m and n (see Table 4.1). When describing the form of nuclear gamma-resonance spectra, it is very important to take into account the atoms of not only the first but also the second coordination sphere.

When the thickness of the layers A and B is considerably greater than the thickness of the interlayer boundaries (according to the results of simulation, the thickness of the interlayer boundaries is equal to 2÷3 monolayers) and, in addition to this, the

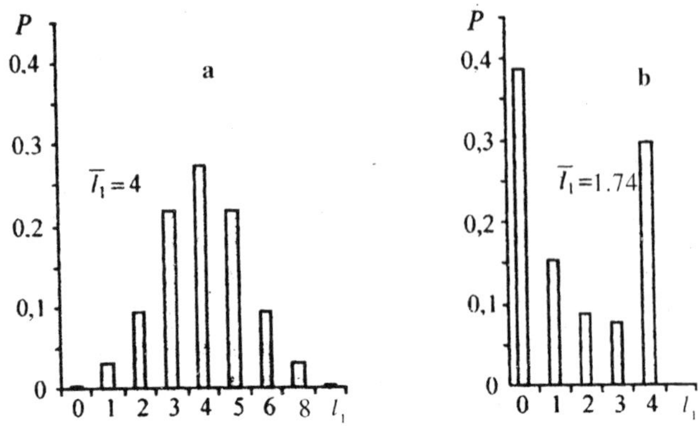

Fig. 4.6. Probabilities of presence in the vicinity of Fe atom of l_1 atoms of second component $B(X)$; a) solid solution $Fe_{50}Cr_{50}$; b) interface Fe/Cr.

shortage (or excess) of the atoms A and B at all interlayer boundaries is the same, the probabilities $P(l_1, l_2)$ are determined by the values of fractions of d_A and d_B ($d_A + d_B = 1$) of the number of deposited layers. The values of these probabilities change with the change of the ratio of the fractions of d_A and d_B but the 'saddle-shape' distribution of probabilities (Fig. 4.6) is unchanged. If the thickness of the layers of the deposited components A and B is specified by arbitrary numbers, the numbers $P(l_1, l_2)$ are obtained by averging out (by the results of deposition of a large number of layers). The type of resultant distribution remains qualitatively the same.

4.9 ON THE POSSIBILITIES OF EXAMINING THE CHARACTER OF DISTRIBUTION OF INTERSTITIAL IMPURITIES

The high sensitivity of the Mössbauer effect to the local atomic environment makes it possible to examine the distribution and charged state of interstitial impurities. The presence of an interstitial impurity in the vicinity of a resonant nucleus is usually manifested because of the local distortion of symmetry (which for the majority of metals, alloys and metallic compounds is usually not lower than cubic). Distortion of the symmetry is accompanied by the formation, in the composition of the Mössbauer spectrum, of components characterised by significant quadropole splitting.

The Mössbauer effect makes it possible to determine the areas of localisation of implanted atoms. In different materials different

Table 4.1 Example of calculated probabilities $P(l_1, l_2)$ for a multilayer sublattice $\left\{A_{S_1+0.7}/B_{S_2+0.3}\right\}_k$ (where S_1 and S_2 are integers)

l_1/l_2	0	1	2	3	4	5	6
0	0.139	0.238	0.009	0	0	0	0
1	0	0.155	0	0	0	0	0
2	0	0.088	0.001	0	0	0	0
3	0	0.049	0.023	0.002	0	0	0
4	0	0.102	0.118	0.049	0.016	0.011	0
5	0	0	0	0	0	0	0
6	0	0	0	0	0	0	0
7	0	0	0	0	0	0	0
8	0	0	0	0	0	0	0

positions are favourable in terms of energy because of specifics of the interaction of external *s*-, *p*-, *d*-, *f*-electrons of the matrix and the impurity.

Parameters such as quadrapole splitting and relative intensity of quadropole components enable in many cases the identification of the interstitial positions (the points of localisation) of the impurity atoms and also examination of their distribution in these positions. A class of permissible models is usually constructed for this purpose. The calculation of the relative intensity of the 'quadropole' component or several components, corresponding to the atoms of the resonant isotope, with interstitial impurities in the vicinity of these atoms, is carried out taking into account the concentration of impurities in tetra- or octapores and special features of the distrbution in these pores (random or ordered) as in the case described in detail for substitutional impurities (paragraph 4.3). The quadrapole splitting may be described, for example, in terms of the approximation of point charges. Different theoretical models are compared with experiment (theoretical and experimental spectra are juxtaposed) on the basis of the statistical criteria of agreement and the optimum model is selected.

In some cases, the separation of a partial subspectrum, associated with interstitial impurities, is carried out using the method of difference spectra (paragraph 5.2 and Appendix 1). This approach has advantages because it excludes any need for using a priori information required for constructing physical models in the first stage of analysis of the spectra, and enables determination of the type of 'quadropole' components whose intensity is often low.

Examples of the Mössbauer studies of the distribution of interstitial impurities in crystal lattices of the alloys are presented in paragraph 5.2.

Chapter 5

MÖSSBAUER STUDIES OF THE ATOMIC AND MAGNETIC STRUCTURE OF ALLOYS

5.1. INVESTIGATION OF THE LONG-RANGE ATOMIC ORDER IN SUBSTITUTION ALLOYS

The classical methods of investigation of the periodic atomic and magnetic structure of solids are the methods based on the diffraction phenomenon, such as x-ray, electron and neutron diffraction analysis, reflecting the regular structure of crystals in the reciprocal space (the method of magnetic neutron diffraction analysis makes it possible to examine the periodic magnetic structure of magnetically ordered crystals). In the second half of the 20th century methods were developed for visualisation of the structure of crystals in the direct space with atomic resolution. In this case, the methods such as electron, field emission, tunnelling and atomic power microscopy, each of which has its advantages and shortcomings are meant. Work was also carried out on the parallel development of an independent group of local–nuclear methods sensitive to the local electron and atom environment of the resonant nuclei.

One such method, highly effective in the examination of the atomic and magnetic structure of solids, was found to be the Mössbauer effect. However, this method was not immediately regarded as an independent method of determination of the parameters of the atomic and local magnetic structure of crystals.

Analysis, carried out in the chapters 2, 3 and 4 of the present book, provides additional justification for the application of the Mössbauer effect for these applications together with the classical methods. This relates to the coupling equations of the parameters of the Mössbauer spectra with the parameters of the short- and long-range atomic order, confirmation of the equivalence of the 'probability' and microdomain models of atomic ordering, and also for the coupling equations of the effective magnetic field on the nuclei of resonant atoms with the values of the magnetic moments of the atoms of the matrix and the impurity.

At present, it may be claimed that, on the whole, there is a good agreement between the results of classical methods and the results obtained using the Mössbauer effect. This is confirmed, in particular, by the results of examination of the long- and short-range atomic order, presented in this chapter.

Of course, it must be remembered that the application of several mutually complementary experimental methods in the examination of the atomic and magnetic structure of the solids makes it possible to obtain the complete and most reliable information on the short- and long-range atomic order.

Consider the examples of application of the Mössbauer effect in investigation of the distribution of atoms in the alloys.

In the alloys examined extensively and efficiently by Mössbauer spectroscopy, it is important to mention solid solutions of aluminium and silicon in iron. The ordering energy of these alloys is very high and, consequently, the formation of the short-range order in this alloy should already be expected for diluted solid solutions (less than 10 at.% Al, Si). In the alloys containing more than 10–15 at.% Al and Si, a long-range order forms which is often difficult to remove even by quenching from high temperatures.

It should be mentioned that the examination of the short-range order in the diluted solid solutions of the substitution type using diffraction methods is complicated because of the very low intensity of diffusion maxima.

Fe–Al alloys were investigated using different methods by a large number of investigators but a number of problems related to the transformations in heating and cooling has not been explained. The phase diagrams of the iron–aluminium system, constructed on the basis of these investigations, greatly differ from each other.

At room temperature, there are three concentration regions: 1) from 0 to 18 at.% Al – α-solid solution, with the BCC lattice, 2) more than 32 at.% aluminium – β-solid solution based on iron monoaluminide with the *B*2 lattice, 3) the intermediate region from 18 to 32 at.% Al containing several phases (Fe_3Al, FeAl, and according to some data, also $Fe_{13}Al_3$). The phase transformations in this region, the type of ordering and the boundaries of existence of different phases are still the subject of discussion. There is no united view regarding the ordering mechanism and also the mechanisms of transformation of one type of ordered structures to another.

A significant contribution to the examination of phase transitions in iron–aluminium alloys has been provided by investigations by the nuclear gamma resonance method. They include mainly Mössbauer investigations of ordered alloys with the composition close to the stoichiometric composition Fe_3Al. The superstructure Fe_3Al (type DO_3) is an ordered structure of the second rank (chapter 4, Figs. 4.2, 4.3). Its formation may be described as the division of the BCC lattice into two simple cubic lattices and the division of the individual simple cubic lattices into two FCC sublattices with the double spacing. In this case, the atoms of aluminium occupy one of the face-centered cubic sublattices, and atoms of iron the three others. However, if the atoms of aluminium occupy one of the simple cubic sublattices, and the atoms of iron the other one, the superstructure of the first rank FeAl forms (type *B*2).

Of course, in the case of binary and multicomponent alloys with a nonstoichiometric composition we can talk about only the preferential population of the sites of specific sublattices because of the shortage of atoms of the one or the other type.

The method of nuclear gamma resonance was used for the first time for examining the superstructure of Fe_3Al (DO_3) in an alloy with a stoichiometric composition [83]. It was established that the atoms of iron in the sublattices *A* (one sublattice) and *D* (combines the three remaining sublattices) have two corresponding sextets of the lines in the spectrum of the alloy, displaced in relation to each other and greatly differing in the strength of the effective magnetic field: at room temperature, respectively, 229 ± 10 and 299 ± 10 kOe.

The differences in the charge and spin density of the electrons on the nuclei of the iron atoms of different sublattices are determined by their different environment: the atoms of the sublattice *A* contain in the first coordination sphere four iron atoms and four aluminium atoms, the atoms of the sublattice *D* contain eight iron atoms (see Fig. 4.2, the sites *A* correspond to sites B_2, and sites *D* to the sites B_1).

The Mössbauer examination of ordering in cooling of an Fe+ 24.3 at.% Al alloy, was carried out by Cser *et al* [84]. At temperatures above 650°C the alloy is disordered and has a corresponding Mössbauer spectrum in the form of a single line (Fig. 5.1). In cooling, a sextet of lines appears in the spectrum and corresponds to the presence of an ordered magnetic phase. At room temperature, the absorption spectrum is represented by the superposition of two sextets, corresponding to nuclei of ^{57}Fe in the sublattices *A* and *D*. On the basis of these investigations, the authors relate the *A*2 → *B*2 transformation to the phase transition of the first kind, and not the second kind, as it was assumed previously.

Analysis of the results of examination of Fe–Al alloys by the method of nuclear gamma resonance was carried out by Lessoile and Gielen [85] who noted the contradictions in these results. These authors investigated alloys with 11–28 at.% Al after quenching from 1000 °C and subsequent annealing at 400 °C for eight days. The alloys with 17.8% and a lower aluminium content remained non-ordered after annealing. The alloy with 19.1 at.% Al after annealing for eight days at 400°C was also disordered, but after 30 days of annealing at 250°C the spectrum of this alloy differed from the spectrum of the non-ordered alloy. The authors assume the existence of an $Fe_{13}Al_3$ superstructure whose formation was assumed in earlier studies, but they think that the data that are available are not sufficient to confirm this fact. They carried out the most detailed examination of ordering of the alloy close to Fe_3Al alloy as regards composition. The results of their measurements for this alloy coincide with the data of other investigations.

Detailed examination of the Fe +20÷31 at.% Al by the method of nuclear gamma resonance was carried out by Kuz'min and Losievskaja in [86].

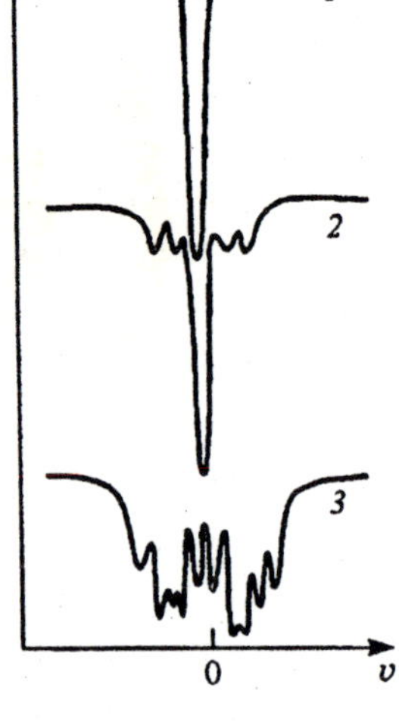

Fig. 5.1 Absorption spectra of Fe + 24.3% at.% alloy at temperatures of: 700 °C (*1*); 550 °C (*2*); 20 °C (*3*) [84].

In [86], the authors determine more accurately the diagram of the Fe–Al system and it was shown that the alloy with the composition Fe_3Al (25.2 at.% Al) is partially disordered. In addition to $P(0)$ and $P(4)$, the absorption spectrum contained components $P(1)$, $P(2)$, $P(3)$ and $P(5)$ ($P(l)$, are the probabilities of atomic configurations, the figure in the brackets is the number of the aluminium atoms closest to iron). Investigating the alloys with the aluminium content smaller than 25 at.%, the authors concluded that there is no need to assume the presence of the $Fe_{13}Al_3$ phase. According to the results, the alloys in the range 20÷25 at.% Al consist of an ordered phase of the type Fe_3Al and the α-solid solution. The authors also concluded that in the investigated alloys with the aluminium concentration higher than 25 at.%, the aluminium atoms populate not only *D* but also *A* sites.

Investigations of an iron alloy with 19.4 at.% Al after low-temperature (300 °C) annealing for 50 hours, carried out by Dorofeev and Litvinov in [5], also did not show the presence of the $Fe_{13}Al_3$ superstructure. Analysis of the absorption spectrum makes it possible to make a conclusion on the existence of the two-phase state of the alloy after such treatment: the α-solid solution with the short-range atomic ordering, and the Fe_3Al phase.

Theory shows that in alloys with a high ordering energy one can expect the formation of superstructures of the third rank with low symmetry. It was noted previously that the data on the existence of the $A_{13}B_2$ superstructure in the Fe–Al alloys are contradictory. As indicated by the results of Mössbauer examination [5], the formation of the superstructure with low symmetry A_5B_3 takes place in annealing of alloys with a high aluminium content.

Annealing of the iron alloys with 40 and 37.5 at.% Al initially quenched at 1000 °C and annealed at 350 °C for 120 hours results in a change in the form of the absorption spectra [5] (the width of the spectra increases) (Fig. 5.2), indicating the redistribution of the atoms of the components. The determination of the type and degree of the long-range order in the alloy with 37.5 at.% Al was carried out using the method proposed in [68]. The type of order was given by the formula $Fe_{16-k}Al_k$, where $k = 1, 2, \ldots, 8$. Figure 5.3a shows the dependence of the probabilities of the configurations of the environment $P^k(l_1, \eta)$ for $k = 4$, 6 and 8 because for the alloys of the examined composition the following types of ordering are most likely to occur: A_3B, A_5B_3 and AB (l_1 is the number of aluminium atoms in the first coordination sphere of the iron atoms; η is the Bragg–Williams long-range order parameter). To avoid overloading

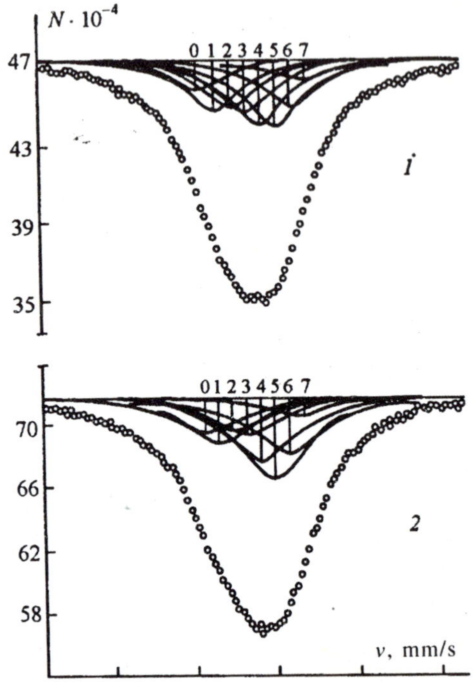

Fig. 5.2 Absorption spectra of iron alloy with 37.5% at.% Al after quenching from 1000 °C (*1*) and annealing at 350 °C, 120 h (2) (thin lines – components of the spectra) [5].

the graph, dependences $P^k(l_1,\eta)$ for all values of η are given only for the A_5B_3 structure.

The Mössbauer spectrum was approximated by the superposition of the lines (see equation (2.7)), the probabilities $P^k(l_1,\eta)$, corresponding to different values of k, l_1 and the long-range order parameter η, were given by an equation similar to (4.9).

The variation of the residual sum of the squares in relation to the degree of order for the alloy with 37 at.% Al, quenched from a temperature of 1000°C, is presented in Fig. 5.3*b* (1). All the curves are characterised by a minimum. The lowest minimum for the quenched alloy corresponds to $k = 8$ and for the annealed alloy (Fig. 5.3*b* (2)) to $k = 6$. Thus, as a result of quenching from 1000°C, the alloy is characterised by the formation of ordering of type *AB* (*B*2) ($\eta = 0.4$). Annealing of the alloy at 350 °C results in the formation of the superstructure of the third rank, described by the formula A_5B_3 (the degree of the long-range order is 0.6). This may also be said of the alloy with 40 at% Al.

It should be mentioned that the concentration of 37.5 at.% Al

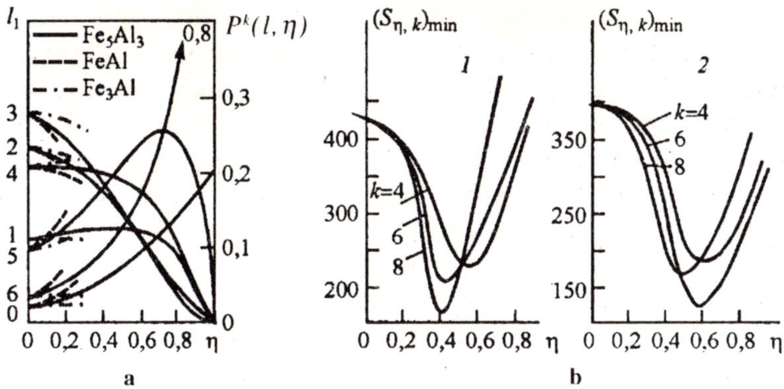

Fig. 5.3 Dependences of *a*) probabilities of atomic environment in an iron alloy with 37.5% at. Al and *b*) residual sums of the squares of the parameter of the long-range order: for the quenched (*1*) and annealed (*2*) specimens [5].

corresponds accurately to the stoichiometric composition A_5B_3. With full ordering of such an alloy (η = 1) there are only two configurations of the nearest environment of the Fe atom with l_1 = 0 and l_1 = 6 adjacent Al atoms. The ratio of the probabilities of these configurations $P(0)/P(6)$ = 1/4. The author is not aware of any examples of similar structures observed by diffraction methods.

Thus, the results of Mössbauer examination testify that the Fe–Al system does not show the formation of low symmetry superstructures on the basis of the α-solid solution at low aluminium concentrations. At the same time, the superstructure of the third rank Fe_5Al_3 forms at a high aluminium content on the basis of the β-phase.

Iron silicides and the solid solution of silicon in iron (α-phase) were investigated by different authors already in the initial stage of development of Mössbauer spectroscopy, simultaneously with iron aluminides and the α-phase of the Fe–Al system. These two systems are similar to each other, and the results of examination of these two systems are usually compared.

The solid solution of silicon in iron is important as the basis of electrical steels. The drop of the plasticity of these steels with an increase of the silicon content is a highly undesirable phenomenon. The low plasticity of iron–silicon alloys already at a relatively low silicon content (3÷4 wt%), like the anomalies of other physical properties, have been explained by some investigators by the partial ordering of the alloys [87].

The compound Fe_3Si has the DO_3 structure. The two non-equivalent positions of ^{57}Fe with different numbers of the atoms of iron and aluminium in the first coordination sphere: 1. eight atoms of iron and 2) four atoms of iron and four atoms of silicon, are linked with the Zeeman sextets of the lines with the effective magnetic field strength of 305 and 195 kOe [88]. The relative shift of the sextet with the weaker field is +0.2 mm/s. Comparison of these data with the results obtained for Fe_3Al intermetallic compound shows that silicon causes a larger change (in comparison with aluminium) in the electron density on the adjacent iron atom nuclei.

In the first Mössbauer investigations of α-alloys with a low silicon content, no formation of a short- or long-range order was revealed. According to the data obtained by Stearns [89], in the iron alloys with up to 10 at.% Si the distribution of the silicon atoms after annealing at 900 °C (2 h) + 400°C (2 h) is close to the statistically uniform distribution. Further examination showed a short-range order in an alloy with 6.5 at% Si after annealing for three hours at 570°C [90]. The authors compared the experimental Mössbauer spectrum with the theoretical spectrum, constructed assuming the statistically uniform distribution of the atoms in the alloy, and established that they greatly differ.

Papadimitrou and Genin [91] compared the intensity of the components Fe_0, Fe_1 and Fe_2 of the experimental spectrum of absorption of a carbon steel with 7 at.% Si (annealing for 3 hours at 750°C) with the probabilities of configurations $P(0)$, $P(1)$ and $P(2)$ in different superstructures of the Fe–Si system, and drew a conclusion on the partial ordering of the steel in accordance with the type $Fe_{15}Si$. The authors show that the possibility of the formation of this superstructure was already predicted in the 30s of the 20th century, but this was not confirmed by experiments because the sensitivity of the diffraction methods was insufficient for the detection of the structures with lower symmetry; the intensity of superstructure reflections of $Fe_{15}Si$ is several orders of magnitude lower than the intensity of the main reflections.

The nature of distribution of the atoms in the α-alloys of the Fe–Si system was investigated in [68–70]. In these studies the authors propose for the first time a method for the determination of the type and degree of the long-range order in solid solutions by means of the Mössbauer effect, as described in chapter 4. Investigations were carried out on alloys with 0.40; 0.85; 1.86; 3.22; 5.92; 6.42 and 7.43 % Si (0.79; 1.70; 3.68; 6.25; 11.12; 12.00 and 13.5 at.% Si, respectively) after annealing for 2 hours at 750°C (cooling in the furnace). They took into account the possibility of the formation of superstructures of the second and third rank $Fe_{16-k}Si_k$ (k = 1, 2, 3, 4).

The calculation of the Mössbauer spectra of the alloys with 0.79; 1.70 and 3.69 at.% Si showed that the minimum of the residual sum of the squares $S_{k,\eta}$ corresponds to the value $\eta = 0$ (there is no long-range order). In this case, the experimental spectra are well described assuming the binomial law of distribution of the probabilities of surrounding the nuclei of ^{57}Fe by different numbers of silicon atoms. In an alloy with 6.25 at% Si, the distribution of the silicon atoms in the crystal lattice sites is ordered. In this case, for k = 2, 3 and 4, the residual sum of the squares $S_{k,\eta}$ continuously increases with increase of η, whereas in the case of k = 1 it has a minimum at $\eta = 0.68$ (similar results for a steel with 4.19% Si are shown in Fig. 5.4). Thus, annealing and slow cooling of the alloy result in the formation of ordering of the type $Fe_{15}Si$ with a high degree of the long-range order.

Investigations of the alloys with 11.10, 12.00 and 13.50 at.% Si using the same procedure (the only difference was that the parameter η was not given with a step but was varied automatically during the search for the extremum of $S_{k,\eta}$; in addition to this, the non-additivity of the contributions of the silicon atoms to the effective magnetic field was taken into account) indicate that they are characterised by an ordering of the type Fe_7Si. The degree of the long-range order is 0.50; 0.56 and 0.54, respectively.

Thus, in the alloys with the silicon content up to 13.5 at.%, two

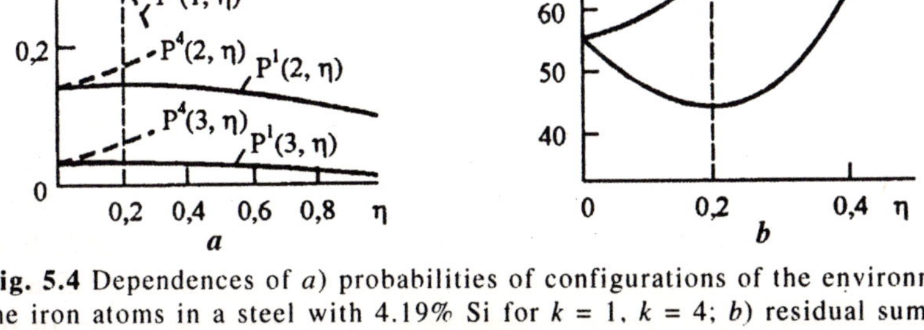

Fig. 5.4 Dependences of *a*) probabilities of configurations of the environment of the iron atoms in a steel with 4.19% Si for k = 1, k = 4; *b*) residual sum of the squares (1 – k = 4; 2 – k = 1) of the long range order parameter [5, 68].

superstructures of the third rank form: $Fe_{15}Si$ and Fe_7Si. The superstructure, whose stoichiometric composition is closer to the composition of the alloy, is the one that forms.

Of considerable interest is the formation of the long-range order in industrial silicon steels after standard treatment [68]. The steel with 6.45% (12.00 at.%) Si is characterised by the ordering of the type Fe_7Si, and the degree of ordering is relatively high (η = 0.58). In a steel with 4.19% Si the ordering of the type $Fe_{15}Si$ (η = 0 .20) is detected only in the specimens showing low plasticity in the bend test (Fig. 5.4). In the specimens of the same steel with increased plasticity, the distribution of the atoms is close to statistically uniform.

The ordering of the type DO_3 forms in alloys with the silicon concentration higher than 13.5 at.%. Figure 5.5 shows the absorption spectra of the iron alloys with 15, 20 and 25 at.% Si after annealing for 2 hours at 750 °C. In the spectrum of the alloy with 25 at.% Si, the two main Zeeman sextets of the lines correspond to the iron atoms with l_1 = 0 and l_1 = 4 atoms of silicon in the first coordination sphere (see Fig. 4.2), the third sextet (its outer lines are indicated by arrows) corresponds to the iron atoms with three silicon atoms in the nearest environment. The strength of the effective magnetic field for the three given configurations of the nearest environment is 310, 201 and 246±5 kOe. The degree of the long-range order of this alloy is relatively high (η = 0.94±0.05) but differs from unity. This explains the presence of a third sextet

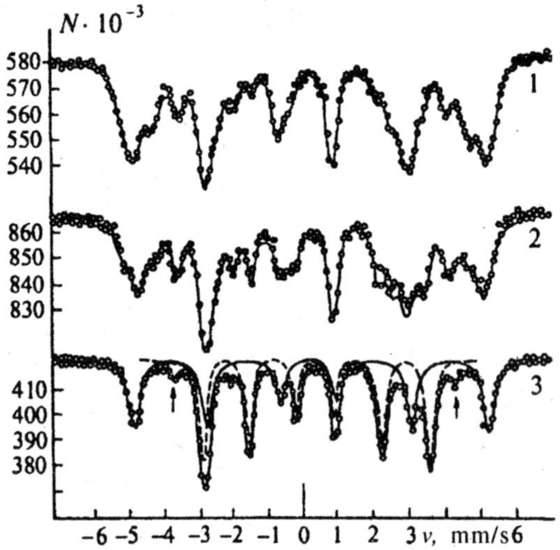

Fig. 5.5. Absorption spectra of iron alloys with silicon: 15 – 1; 20 – 2; 25 – 3 (solid lines are theoretical, components of the spectra are given for Fe_3Si) [5, 79].

with low intensity. The same two main components are also detected in the spectra of the alloys with 15 and 20 at.% Si, but in addition to these components there are also other clearly observed sextets. The deviation from stoichiometry is accompanied by a decrease of the degree of the long-range order ($\eta = 0.61$ and $\eta = 0.69 \pm 0.5$ for the alloys with 15 and 20 at.% Si, respectively).

Additional possibilities in the examination of the special features of the atomic structure of solids by means of the Mössbauer effect are offered by consideration of the magnetic dipole contributions of the impurities in H_{ef} on the ^{57}Fe nucleus. As reported in the paragraphs 3.3.2 and 3.3.5, the mean values of the dipole contributions in the cubic matrices with the random distribution of the impurities are equal to zero. This indicates only widening and the absence of shift of the Mössbauer lines. However, the situation changes in the presence of atomic ordering.

Dorofeev and Litvinov [52] investigated the role of magnetic dipole contributions in the examination of atomic ordering in ferromagnetic alloys with the BCC and FCC lattice. This makes it possible to describe more accurately the special features of Mössbauer spectra and the pattern of atomic ordering. On the basis of the results of these investigations it was established that in silicon iron with 5.7 at.% Si, a directional order forms after slow cooling from 950°C in the distribution of the atoms of iron and silicon. It has been shown that the nuclear gamma resonance spectra of silicon iron are described most efficiently assuming the formation of the order of the $Fe_{15}Si$ type which may be described in terms of the isotropic short-range order in the first coordination sphere and the anisotropic short-range order in the second coordination sphere. In this case, the chains of the atoms of iron and silicon are oriented along the axis of the easiest magnetisation <100>.

The possibility of formation of an anisotropic atomic order in the Fe+6 at.% Si alloy was indicated previously in [92]. This was based on the insufficiently accurate description of the form of Mössbauer spectra in the model of the isotropic short-range order.

In [53], on the basis of high-temperature investigations and modelling of the gradient of the electrical field (GEF), the authors determine the orientation of the effective magnetic field and main axes of the GEF in *a*- and *b*-modifications of iron monoboride FeB. A model of an α-FeB magnetic structure was proposed. The experimental results show that the difference of the values of H_{ef} in *a*- and *b*-FeB is determined by the anisotropic contribution. Estimates were obtained of the anisotropic and isotropic contributions to the value of H_{ef} for both phases, which were equal to 10.9 and 2.7 T (109 and 27 kOe), respectively.

The principles and methods of Mössbauer analysis of the atomic and magnetic structure of the solids, described in the chapters 2–4 and in the present chapter, are employed widely in investigations at the present time. The distinctive relationship between the composition of the nearest atomic environment and the magnetic state of the resonant isotope in magnetically ordered crystals facilitates and widens the possibilities of examination of both the atomic and magnetic structure of solids by means of the Mössbauer effect.

In many cases, the Mössbauer effect is the only method which makes it possible to investigate the formation of the short- and long-range atomic orders in solid solutions (identify the type of structures formed and determine the degree of their perfection in examination of the long-range order). In other cases, the application of the method is more efficient in comparison with the conventional methods. This relates to the examination of diluted solid solutions, ordered structures of the high rank ($A_{15}B$, A_7B, A_5B_3, ...), the initial stages of atomic ordering when the regions of coherent scattering are small (<8–10 nm), and the sites of the reciprocal lattice do not form, alloys with close electronic factors of scattering of the components (closely spaced in the periodic table of elements). This also refers to the investigations of multi-component alloys where the determination of a large number of independent probabilities P_{ij} of the population of the sites of sublattices (or order parameters η_{ij} linked with them by a linear relationship) requires labour-consuming diffraction investigations using several types of radiation with different wavelengths [93].

Because of its universal nature: the possibility of simultaneous examination of the atomic and magnetic structure of condensed media, the Mössbauer effect is utilised in the selection of optimum alloying and conditions of heat treatment of magnetic materials.

FeAl alloys with 24 and 25 at.% Al (containing, according to the literature data, different amounts of $B2$, DO_3, and non-ordered α-phase) are interesting both as model objects (solid solutions of the transition metal and sp-element) and also as magnetic and structural materials. In [94], the Mössbauer effect was utilised for the determination of the effect of the concentration of aluminium and conditions of heat treatment on the local atomic structure and phase composition of these alloys. It has been shown that the mean effective magnetic field $\langle H \rangle$, which correlates with saturation magnetisation, depends on the mean cooling rate during precipitation of the respective phases.

Experience with the application of Mössbauer spectroscopy for the examination of the long-range atomic order in binary systems is used widely in the investigations of the preferential population of sublattice sites by the atoms of components in ternary and multi-component alloys.

The effect of heat treatment on the properties of ferromagnetic films $Fe_{71.3}Al_{9.8}Si_{15.9}$ with a thickness of 1 μm, produced by deposition on glass substrates, was investigated in [95]. The films were annealed at 300÷700°C. Examination by the Mössbauer effect, x-ray diffraction and magnetic methods showed that annealing at 500°C results in the formation of the well ordered DO_3 structures in the films. These films are characterised by the minimum coercive force (0.2÷0.3 Oe). Annealing in the temperature range 600÷700°C results in the partial failure of the DO_3 structure with the formation of phases with a variable composition.

The authors of [96] carried out Mössbauer investigations of quasi-binary alloys of the $Fe_{0.75-x}Mn_xAl_{0.25}$ system using the method of restoration of the distribution functions of magnetic hyperfine fields. The experimental results show that the preferential substitution of Fe sites by manganese in one of the sublattices results in the ordering of the ternary alloy according to type *B*2. Examination of the kinetics of ordering in $Fe_{0.45}Mn_{0.30}Al_{0.25}$ alloy showed large differences in the relaxation times, characterising ordering in different sublattices.

The $Fe_{3-x}Cr_xAl$ ($x < 0.6$) alloys were investigated by the methods of Mössbauer spectroscopy, x-ray diffraction and neutron diffraction analysis [97] in the temperature range 10–300 K. At all investigated concentrations of chromium the single-phase state with the DO_3 structure is retained. It has been established that the chromium atoms occupy mainly the sites *B* of the crystal lattice (sites B_1 in Fig. 4.2). The lattice constant decreases with increase of the chromium concentration. The values of magnetic moments of iron in different crystallographic positions and the effect of the nearest atoms of chromium and aluminium on the values of H_{ef} and μ_{Fe} were estimated.

The Mössbauer effect was also used for the examination of the process of breakdown of freshly quenched Fe–Co–Al alloys with the formation of the *A*2+*B*2 structure [98]. The Mössbauer spectra are divided into components, corresponding to different nearest environment of the Fe atom (for the *A*2 and *B*2 phases), using the empirical relationship between the value of the effective magnetic field and the number of the cobalt and aluminium atoms in the nearest environment of the atom of the resonant isotope. The boundaries of the phase diagram for the aluminium-rich and depleted sides were determined.

In [99], using the methods of nuclear magnetic and nuclear gamma resonance and also neutron diffraction it was shown that the structure of the Ce_2F_{17} compound is close to the ideal structure of the type Th_2Zn_{17}. In the same study it was also shown that the intermetallic compound Lu_2Fe_{17} represents the disordered variant of the structure Th_2Ni_{17}, characterised by the exchange of sites between the rare-earth metal in 2*b* positions and 'dumbbells' of Fe (in 4*f* positions), and also by partial substitution of the 4*e* positions by the iron atoms. In addition to this, a certain amount of Y is introduced into the 2*c* sites, causing distortion and splitting of the 12*j* position of Fe into two positions. The results of examination of neutron diffraction, which also shows the distortion of the lattice, are in good agreement with the data of nuclear magnetic and nuclear gamma resonance.

In [100], the Mössbauer effect was used to examine the process of nanocrystallisation of the amorphous alloy $Fe_{73.5}CuNb_3Si_{17.5}B_5$ at 490°C. Two stages of the process were detected. In the first stage, there are major changes in the parameter of the short-range chemical order in the amorphous phase. The second stage is characterised by the formation of nanocrystals with the DO_3 structure. These results are supported by the data of differential scanning calorimetry and transmission electron microscopy. The results show unambiguously that the process of nanocrystallisation takes place by the nucleation and growth of nuclei. It should be stressed that the possibility of identification of the formation of DO_3 structure in nanocrystalline volumes of the alloy indicates the obvious advantages of the Mössbauer effect in comparison with the traditional methods.

The susceptibility to ordering in many alloys is so strong that the formation of the long-range order cannot be suppressed even by quenching from high temperatures, and also by ion mixing (see chapter 6). The effect of disordering (disruption of chemical and, in some cases, also topological order) can be achieved only by mechano-chemical methods, subjecting materials to high-intensity plastic deformation at reduced temperatures, for example cold rolling, shear under compression, or milling in a ball planetary mill. This experiments are interesting both from the purely scientific viewpoint – the formation of disordered objects for the examination of chemical bonding (without superposition of the effects of atomic ordering) and from the practical viewpoint – the dissolution of the ordered phases by mechanochemical methods, production of materials with special mechanical, electrical and magnetic properties.

In studies carried out by Sagaradze, Shabashov, *et al*, the method of Mössbauer spectroscopy was used for the detection of the processes of dissolution of ordered intermetallic [101, 102] and carbide [103] phases in Fe–Ni austenite under the effect of high-intensity cold plastic deformation.

For the quantitative analysis of the kinetics of the process of dissolution of intermetallics Ni_3Ti [101, 102], Ni_3Al, Ni_3Si, Ni_3Zr [102] investigators constructed the calibration dependence of the mean field $\langle H \rangle$ on the nickel content in the binary austenitic alloys of iron with 29.5–39.9 at.% Ni. Deformation was carried out by rolling in smooth rolls ($e \leq 4.23$) and by shearing under compression ($e \leq 5.93$). On the basis of analysis of the type of functions of the distribution of hyperfine fields $P(H)$ and the variation of $\langle H \rangle$ in deformation it was established that intensive plastic deformation results in the complete dissolution of intermetallic particles (spherical particles of the FCC γ-phase with a diameter of 5 nm and large plates of the HCP η-phase with a diameter of 50 nm), as a result of drift of the atoms of intermetallics in the field of the stresses of climbing dislocations. The results show that the nickel concentration in the FCC γ-matrix depends in a linear manner on the degree of deformation. The critical degree of deformation for the start of dissolution of the particles is $e_{cr} \approx 0$ for fine particles with a diameter of 5÷6 nm and $e_{cr} \approx 0.5$ for large particles, with a diameter of 50 nm.

The high sensitivity of the method of nuclear gamma resonance made it possible to detect, on the basis of the type of function $P(H)$ [101], the inhomogeneity of the distribution of nickel in the FCC matrix after the deformation dissolution of intermetallic compounds (short-range atomic order). The degree of homogeneity of the solid solution increases with increase of the degree of deformation.

Elsukov *et al* [104–118] investigated the structure and magnetic properties of ordered and also maximally disordered (produced by electrochemical and mechanochemical methods) micro- and nano-crystalline amorphous alloys of iron with *sp*-elements (Al, Si, P, Sn). These alloys are suitable modelling objects (with the properties similar to the amorphous multi-component alloys on which they are based). It is possible to form in them the single-phase state with the BCC lattice in a wide range of the concentrations of the *sp*-element. They are susceptible to the formation of ordered structures (usually of the type $B2$ or DO_3, and also superstructures of the third rank $A_{15}B$, A_7B, A_5B_3, etc [5, 68, 91], at the content of the *sp*-element larger than 10÷15 at.%).

The use of disordered specimens of alloys with the content of *sp*-elements up to 50 or more at.% makes it possible to compare the concentration dependence of the parameters of electrical and magnetic hyperfine interactions (without superposition of the effects associated with the formation of the long-range atomic order). Mechanoactivation was carried out by milling of the alloys, melted by the conventional methods, in a planetary ball mill.

The authors managed to produce iron alloys with all the above *sp*-elements in the concentration range from 0 to 50÷70 at.% in the disordered state (the results of Mössbauer and x-ray diffraction investigations indicate the absence of the long-range atomic order in the mechanically activated specimens). The non-ordered specimens were used for the measurement of the concentration dependences of the parameter of the crystal lattice, saturation magnetisation and the parameters of the electrical and magnetic hyperfine interactions of ^{57}Fe nuclei in the previously mentioned range of the concentrations of the *sp*-elements.

On the basis of the data, obtained by the methods of Mössbauer [101–113, 118] and EXAFS spectroscopy [114, 117] and also using the method of x-ray diffraction analysis, it was concluded that the entire range of the existence of the BCC structure may be divided into concentration ranges with 'good' at $c < c_1$ and 'poor' ($c > c_2$) BCC structures [104], divided by the transition region ($c_1 \leq c \leq c_2$). The 'good' structure is the structure with the random distribution of the atoms of iron and the *sp*-element in the nearest environment of the ^{57}Fe nuclei. The probabilities of the local atomic environment of ^{57}Fe and the intensity of the appropriate components of the Mössbauer spectra are described in this case by a binomial distribution with the coordination number $z = 8$ (the authors justified the sufficient accuracy of the model taking into account only the first coordination sphere). For the region ($c < c_1$) we can ignore the formation of the pairs of *sp*-atoms at the distance of the nearest neighbours and the appropriate local distortions of the BCC lattice.

In the concentration region of the 'poor' BCC structure, the local distribution of the atoms, according to the conclusions by the authors, cannot be described using the value $z = 8$ for the first coordination sphere. The attempts to describe the Mössbauer spectra on the assumption of the presence of the short-range atomic order result in the values of the parameter of the short-range order α_1 (in the first coordination sphere) considerably higher than the maximum permissible values. The authors concluded that the best description of the Mössbauer spectra for the range of the concentrations from c_1 to c_2 (c_2 and c_1 are individual for every *sp*-element) is obtained assuming the random distribution of the coordination numbers for ^{57}Fe (in the range $z = 8 \div 14$) at the random distribution of the atoms for every value of z. In the case with $c > c_2$, the best description of the curves of resonance absorption is obtained at $z = 14$. The 'poor' BCC structure is characterised by the greatly increased width of the structural reflections. This is valid both for the systems FeAl, FeSi and FeSn with the BCC lattice, and for the FeP system in the amorphous–crystalline state. At $c > c_2 \approx 20$ at.% FeP alloys are characterised by the formation of the amorphous state. In the FeSi alloys with $c > c_2 \approx 33$ at.% there can be realised both the crystalline and amorphous state (at the evaporation of the components on the substrate, cooled by liquid nitrogen [119, 120]). For the 'poor' BCC structure, it is necessary to take into account the existence of the *sp*-atoms in the nearest coordination contact, where the strong interatomic interaction of the *sp*-states may lead, according to the authors, to the situation in which it is not possible to separate the first and second coordination spheres (in contrast to the ideal BCC lattice).

Analysis of the experimental results shows that the convergence of the effect of the first and second coordination spheres in the magnetic sense (with increase of the concentration of *sp*-elements) is indicated by the data of Mössbauer spectroscopy, presented by the authors: the values of the contributions to the hyperfine magnetic field on the nucleus from the atoms of the *sp*-element, distributed in the first and second coordination sphere of ^{57}Fe, become more and more similar. At the same time, the results obtained by x-ray diffraction and EXAFS spectroscopy [114, 117] do not provide an unambiguous confirmation of the 'merger' of the first and second coordination spheres, although the tendency for the convergence of the radii R_{Fe-Si} for the first and second coordination spheres with increase of the concentration from zero to 33 at.% is evidently recorded in the case of the FeSi alloys. For the FeSn alloys this tendency is not detected [117]. At the same time, the Sn atoms are redistributed from the second to the first coordination sphere, indicating short-range ordering.

All these factors make it possible to propose an alternative hypothesis regarding the reason for the detected behaviour of the magnetic characteristics. The convergence of the contributions to H_{ef} from the atoms of the first two coordination spheres may be determined by the increase in the period of oscillations of the spin density of the conduction electrons (see chapter 3, Figs. 3.2, 3.4) as a result of the decrease of the concentration of the conduction electrons and, correspondingly, the Fermi radius k_F. The latter is confirmed by the variations of the experimental concentration dependences of the isomeric shift $\delta(c)$ for the investigated alloys of the *sp*-elements [104]. Finally, this results in the effective (in respect of the magnetic hyperfine structure) coordin-ation number $z_{ef} = z_1 + z_2 = 14$. Taking into account the results obtained in EXAFS spectroscopy [114, 117] it is also not possible to reject completely the role of the short-range order. The inhomogeneity in respect of the composition and degree of atomic order and also the possible presence of microvolumes with the structure of the long-range order may result in the values of the parameters of the short-range order higher than the maximum permissible values.

These considerations do not contradict the experimental data presented in [104–110] and, on the whole, do not affect the statements of the description of the concentration behaviour of the magnetic properties of these alloys proposed by the authors.

Mössbauer investigations of the atomic and magnetic ordering in amorphous Fe–Si alloys were carried out in [119, 120].

The ordered phases with a variable composition may be detected not only in metallic alloys but also in other inorganic compounds. The similarity of the problems of their Mössbauer analysis irrespective of the large differences in the chemical nature becomes clearly evident.

Rusakov *et al* [121–132] developed special methods of Mössbauer spectroscopy of locally inhomogeneous systems (LIS). The presence, in the composition of different substances, of identical or similar crystalline phases, configurations of the local atomic environment of the resonant isotope, structural defects, etc, was the main basis for the development of a systematic approach to the analysis of greatly differing objects with the long-range atomic order, including metallic alloys and compounds.

Investigations were carried out for the crystallochemical identification of partial Mössbauer spectra of nuclei of ^{57}Fe and ^{119}Sn, and the appropriate values of the parameters of the HFI (hyperfine interactions) for the LIS of different nature were obtained, having scientific and applied value: alloys [129–132], minerals [124–128], 'target'–'implanted impurity' systems [122, 123,], etc. Databases and software facilities, in particular, the MSTools package, were developed for the processing and analysis of the Mössbauer data. The application of this package by experts in the area of Mössbauer spectroscopy both in Russia and abroad showed its high efficiency.

5.2. EXAMPLES OF INVESTIGATION OF THE SHORT-RANGE ORDER. INTERATOMIC CORRELATIONS, LOCAL ATOMIC STRUCTURE, NON-IDEAL SOLID SOLUTIONS

The deviation of the distribution of the atoms in alloys and steels, after initial quenching and cold deformation, from the statistically uniform distribution, associated with the formation of the short-range order in them, may occur as a result of heat treatment (heating, isothermal holding and subsequent cooling). Usually, this has a marked effect on the physical and mechanical properties of materials. In some cases, the degree of the short-range order may be high, and the effect on properties insignificant, whereas in other cases there may be large changes in the properties, determined by the formation of the short-range order. We can present a relatively large number of examples of efficient application of the Mössbauer effect for the examination of the short-range order in different systems, including for cases in which the conventional methods are ineffective. The latter relates to the examination of the diluted solid solutions, systems with similar scattering capabilities of the components, and also nanocrystalline (x-ray-amorphous) materials. For example, because of the similar atomic amplitude of scattering of the components, it is very difficult to carry out x-ray diffraction study of the processes of redistribution of atoms in FeCr, FeMn, FeCo and FeNi alloys.

The phase separation of the Fe–Ni (30÷40% Ni) alloys after irradiation with electrons in the range 80÷250°C was investigated in [133]. The formation of the short-range order in the alloys of the same composition as a result of annealing and electron irradiation was investigated by Rodionov *et al* [134]. It was concluded that the type of short-range order depends on the dose and temperature of irradiation.

The interest in the investigation of the short-range order in the Fe–Al alloys is caused mainly by the fact that there are large changes in the properties of the solid solution with the aluminium content of up to 20 at.% at low-temperature (200÷450°C) annealing after quenching or plastic deformation (formation of the *K*-state).

The examination by electron microscopy [135] shows the formation of zones ordered on the basis of the DO_3 type in the Fe + (18÷21.5) at.% Al, quenched from 850°C and annealed at 260–360°C.

Mössbauer investigations were carried out into the Fe +10.4 at.% Al alloy after ageing at 230–300°C (1000 min) (see [5, p.75]). The authors concluded that the *K*-state is not associated with the ordering by the type Fe_3Al, FeAl or $Fe_{13}Al_3$. Brummer *et al* [60] used the method of nuclear gamma resonance for the examination of the redistribution of atoms in annealing (300°C, 10 h) alloys of iron with 12.3 at.% aluminium. The annealed specimens contained a short-range order ($\alpha_1 \approx -0.1$; $\alpha_2 \approx 0.04$). For the quenched and deformed specimens $\alpha_1 = \alpha_2 = 0$. The formation of the short-range order correlates with an increase of electrical resistance. The authors explain this by the increase of the number of Fe–Al bonds as a result of short-range ordering (at $\alpha_1 = -0.1$, the number of bonds increases by 11% in comparison with the number of such bonds in the non-ordered alloy).

Figure 5.6 shows the variation of electrical resistance and also of the parameters of the spectrum of resonance absorption and the Cowley parameter of the short-range order for the Fe +15.3 at% alloy after annealing at 300°C for various periods of time [136]. The variation of all parameters of the Mössbauer spectrum correlates with the variation of electrical resistance. The changes of the parameter of the short-range order for the first coordination sphere of the iron atoms with annealing time have the form of a curve with saturation.

The isomeric shift for the iron nuclei increases with increase of the number of aluminium atoms in the nearest environment of iron. This fact is explained by the authors by the transition of part of the external electrons from the aluminium atoms to the 3*d*-band of iron.

The mean strength of the effective magnetic field $\langle H_{ef} \rangle$ on the iron nuclei decreases with increase of annealing time (Fig. 5.6), which (at $\Delta H_1 < 0$ for Fe–Al alloys) is caused by the increase of the effective concentration of aluminium in the first coordination sphere of the iron atoms, see (4.34).

An abrupt change of the mechanical properties (in particular, impact toughness) was found for Fe–Mn BCC alloys. It is well-known [137] that the martensitic iron–manganese steels show brittleness after tempering in the range of 300÷500°C and also at temperatures above 600°C (tempering for several hours reduces the impact toughness from $12 \div 18 \cdot 10^5$ to $2.4 \cdot 10^5$ kN/m^2). In high-

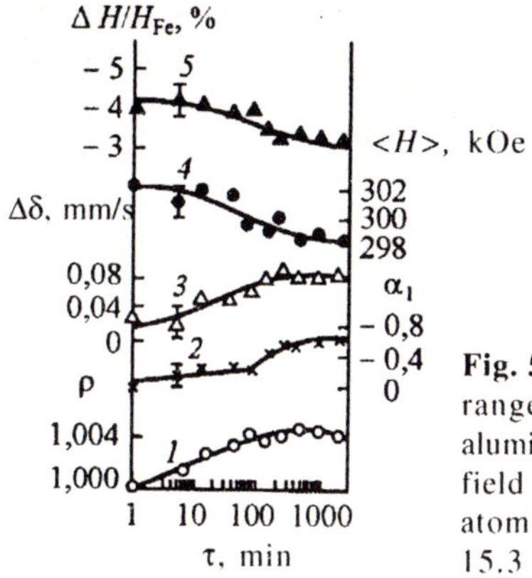

Fig. 5.6 Dependence of electrical resistance (*1*), short-range order parameters (*2*), the contribution of the aluminium atom to isomeric shift (*3*), the mean effective field (*4*) and the relative contribution of the aluminium atom to the hyperfine field (*5*) for an iron alloy with 15.3 at.% Al on annealing time [5,136].

temperature tempering of Fe–8Mn steels (Fe + 8 at% Mn) studies showed the formation of up to 30÷50% austenite unstable in plastic deformation. The low plasticity of the steels after this type of treatment is associated with the transformation of austenite in the course of deformation at loading. The reasons for irreversible temper brittleness of the manganese steels after tempering in the range of 300÷500°C have not been explained. It has been established [137] that the reduction of fracture energy is not accompanied by the variation of the phase composition of the specimens and is not removed by a decrease of the content of interstitial impurities in the alloy.

In the replacement of part of manganese by nickel (total content of nickel and manganese is 8÷9%), the brittleness of the alloys after low temperature tempering decreases, indicating the link of the temper brittleness with the effect of manganese on the properties of the α-solid solution. In holding Fe–8Mn steel in the temperature range 300÷500°C, the formation of austenite is basically possible and this may result in a decrease of impact toughness (when the austenite precipitates at the grain boundaries). However, in actual holding periods (several hours), the occurrence of processes preceding the formation of austenite, associated with the formation of sub-microvolumes, enriched and depleted with manganese, is more likely. Examination of the processes of redistribution of iron and manganese in the investigated alloys by the method of x-ray diffraction is practically impossible because of the close atomic numbers of manganese and iron and, correspondingly, the atomic factors of scattering of these elements.

Figure 5.7 shows the variation of the mechanical properties and

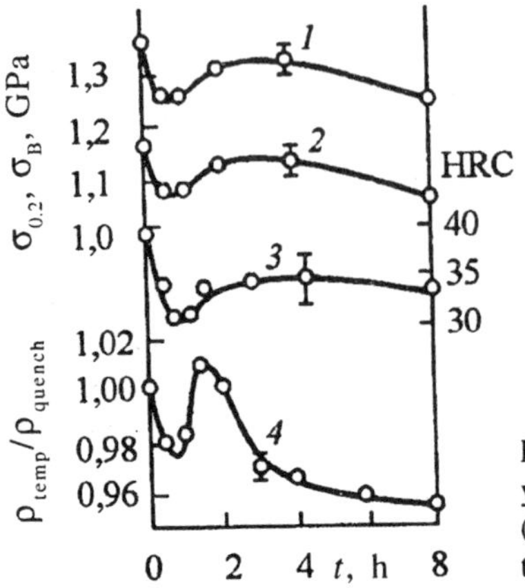

Fig. 5.7 Dependence of ultimate strength (*1*), yield limit (*2*), hardness (*3*) and electrical resistance (*4*) of Fe+8 at.% Mn alloy with 0.07% C on tempering time at 350 °C [5, 138].

electrical resistance of the Fe–8Mn steel as a result of tempering at 350°C, corresponding to the maximum effect of brittleness [138]. The nature of these variations in the initial stages of tempering, and also the fact that no new phases appear in the steel as a result of tempering, make it possible to assume that the anomalies of the properties are associated with the intra-phase processes, possibly with the formation of a short-range order.

The Mössbauer spectra of the alloys are presented in Fig. 5.8. A line of austenite appears in the central part of the spectrum of the alloy with 0.07% C. In tempering, the amount of austenite increases from 3 to 5%. In the alloy with 0.008% C there is no

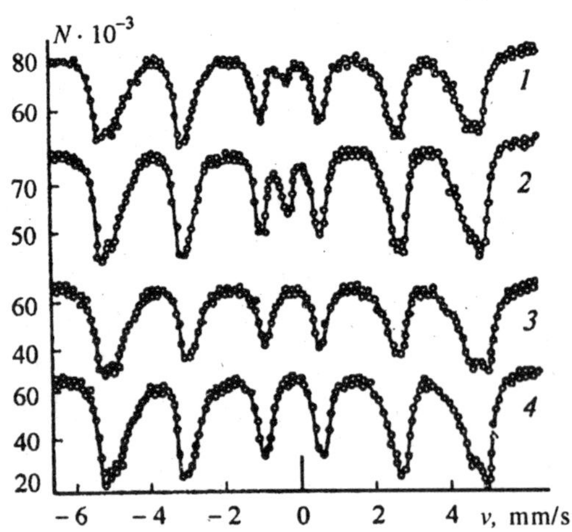

Fig. 5.8 Absorption spectra of iron–manganese alloys with 0.07% C (*1, 2*) and 0.008% C (*3, 4*). Treatment of the specimens: quenching from 800 °C (*1, 3*). quenching from 800 °C and tempering at 350 °C, 10 h (*2, 4*) [5].

austenite in the quenched state or in the state after tempering for 10 hours. Tempering results in the change in the shape of the external peaks of the spectra, i.e. the variation of the ratio of the intensity of the components, corresponding to the resonance nuclei of iron with different numbers of the nearest atoms of the impurity. The spectra of both alloys are characterised by an increase of the intensity of the component, determined by the absorption of the γ-quanta by the iron nuclei with the 'purely iron' environment. The mathematical processing of the spectrum of the single-phase steel was carried out taking into account the effect of remote coordination spheres on the width of the lines using the equation (4.28), which makes it possible to increase the accuracy of determination of the short-range order and other quantities.

The form of the external peaks $-1/2 \rightarrow -3/2$, $+1/2 \rightarrow +3/2$ of the Mössbauer spectrum was defined by the relationship similar to equation (4.38), with the only difference that the probability $P(l_1, \alpha_1)$ of the iron atom being surrounded by different numbers of the manganese atoms for the given value of the parameter of the short-range order α_1 was determined by the relationship (4.12) ($i = 1$).

The calculation of the values of the parameter of the short-range order (in relation to the tempering time) for an alloy with 0.07% C is difficult because the manganese content of the austenite, formed during tempering, is unknown, and the nature of redistribution of the atoms of the components in this alloy may be described only qualitatively. Quantitative analysis of formation of the short-range order was carried out for a single-phase alloy with 0.008% C.

The results of analysis of γ-resonance spectra [138] make it possible to examine the kinetics of intra-phase changes taking place in the iron–manganese austenite during tempering. The comparison of the kinetic curves, describing the variation of the degree of the short-range order α_1 and impact toughness a_n (Fig. 5.9) indicates the existence of a significant correlation between these quantities (correlation coefficient $\rho = -0.85$). The abrupt decrease of impact toughness is detected in the same time period as the abrupt increase of the degree of the short-range order. Processing of the resultant data makes it possible to determine the empirical relationship between a_n and α_1:

$$a_n = a_n^0 - \alpha_1 \frac{a_n^0 - a_n^\infty}{\alpha_1^\infty}, \quad \alpha_1 \le \alpha_1^\infty, \tag{5.1}$$

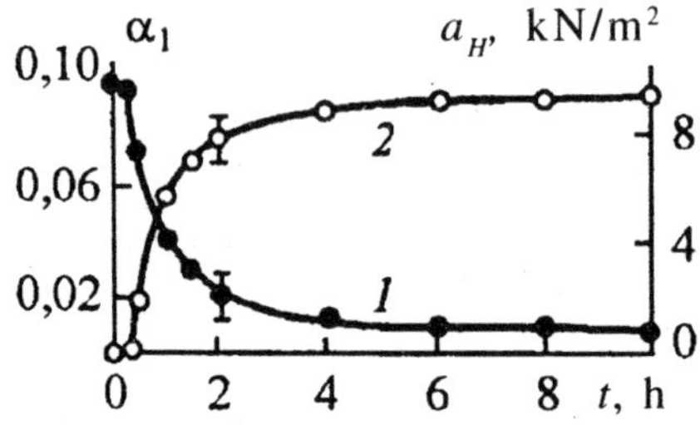

Fig. 5.9 Dependence of impact toughness (*1*) and the short-range order parameter (*2*) of an alloy with 0.008% C on tempering time at 350 °C [5,138].

where the indexes 0 and ∞ correspond to the initial and final values of impact toughness and of the order parameter.

The positive value of parameter α_1 indicates that tempering is accompanied by short-range separation (break-down) of iron–manganese martensite to zones enriched and depleted with manganese. The depend-ence of α_1 on tempering time t has the form of a curve with a latent stage, characteristic of the process of the redistribution of the atoms of the component taking place as a result of the nucleation and growth of nuclei, and is not characteristic of spinodal-type processes.

In order to evaluate the composition of the microvolumes of the alloy, the Mössbauer spectrum of an alloy with 0.008% C, held for 10 hours at 350°C $\left(\alpha_1 \cong \alpha_1^\infty\right)$, was processed using a more complicated model which, in accordance with the assumption on the mechanism of the nucleation and growth of the zones, considers the possibility of separation of the solid solution into microvolumes, enriched and depleted in manganese. The probability of the presence of l_1 atoms of the impurity in the first coordination sphere of the iron atoms was given by the sum of the binomial distributions taking into account the fraction of the iron atoms in the regions enriched and depleted with manganese:

$$P(l_1, c_1, c_2) = n_1 C_8^{l_1} c_1^{l_1} (1-c_1)^{8-l_1} + n_2 C_8^{l_1} c_2^{l_1} (1-c_2)^{8-l_1}, \tag{5.2}$$

where c_1 and c_2 are the manganese concentration in Mn-depleted and Mn-enriched zones; n_1 and n_2 are the fractions of the iron atoms in these zones; $n_1 = [(c_2 - c) \cdot (1 - c_2)]/[(c_2 - c_1)\cdot(1-c)]$; $n_2 = 1 - n_1$; c is the mean manganese concentration of the alloy. Calculations carried out in the same manner as the search for parameter α_1, gave the values of $c_1 = 2.0$, $c_2 = 12.4 \pm 1.0$ at.%.

The formation of sub-microvolumes, enriched with manganese (with increased strength of atomic bonds and, correspondingly, a higher value of the Peierls barrier) should be accompanied by hindering of plastic deformation of the alloy. Since the correlation parameter for α_1 and α_n is close to unity, and no new phases form in the process of tempering of carbon-free martensite, it may be assumed that the cause of formation of brittleness of the investigated Fe–8Mn steel is the short-range separation of the substitution solid solution.

In addition to the results of examination of the hyperfine magnetic structure of the alloys FeAl, FeCr, FeCrCo and FeCo (based on iron), paragraph 5.3.1 also gives the additional data on the formation in these alloys of a short-range atomic order, obtained using the Mössbauer effect.

The formation of the short-range order of the type of short-range separation was detected by nuclear gamma resonance in iron alloys with increased (~20%) manganese content [139]. These two-phase alloys ($\gamma + \varepsilon$) are used as a base of high-strength nonmagnetic materials. The properties of the latter depend on the ratio of the content of the γ- and ε-phases. The reduction of the stability of austenite as a result of alloying is a natural method of hardening these alloys.

Since the ε-phase exists in the Fe–Mn system in a narrow concentration range, i.e. large change of the content of the phase takes place already after adding 1÷2% of alloying elements, and the effect of different impurities differs. It may be assumed that the variation of the stability of the phases in the Fe–20Mn type alloys in alloying is associated with the effect of alloying elements on the nature of distribution of the atoms of the main components.

The absorption spectrum of the type Fe–20Mn alloy (base: Fe + 20 at.% Mn) at room temperature consists of a relatively wide line which is a superposition of the single line of the ε-phase and the slightly resolved Zeeman sextet of the γ-phase (Fig. 5.10). Since the isomeric shifts of the γ- and ε-phases are similar, the spectrum is almost symmetric. With decrease of temperature below

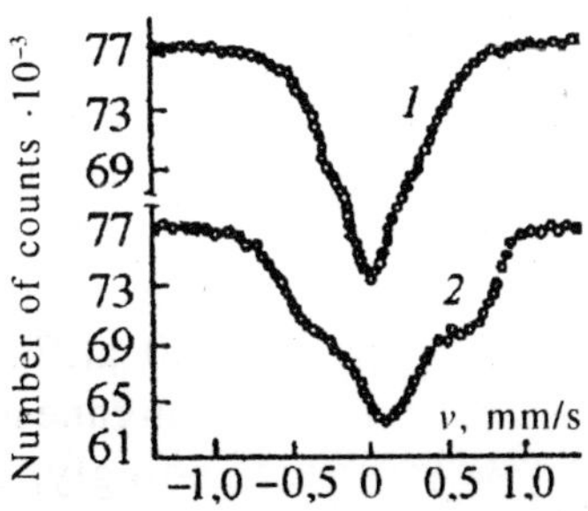

Fig. 5.10 Absorption spectra of Fe–20Mn alloy at 20 (*1*) and −150 °C (*2*).

the Neel point of the ε-phase (for the investigated alloy T_N^{ε} ~-100°C, and T_N^{γ} ~+90 °C), the form of the spectrum changes and the spectrum is described by the superposition of two slightly resolved Zeeman sextets. The extraction of information from spectra of this type (for example, on the nature of distribution of the atoms) by the method used efficiently for the analysis of ferromagnetic alloys and steels, is complicated because of the problems in the separation of components corresponding to the iron atoms with different nearest environment. In the investigated case, it is expedient to use the Window method, described in appendix 1.

Figure 5.11 shows the $P(H)$ functions of the two-phase alloys Fe–16Mn and Fe–21Mn (with 16 and 21 at% Mn), characterising the distribution of the hyperfine magnetic fields in the ε-phase (low-field maximum) and the γ-phase (the maximum at higher values of H_{ef}). The distribution of $P(H)$ also reflects differences in the content of the γ- and ε-phases in these alloys, 55±5% of the ε-phase in the Fe–16Mn alloy and 42±5% in the Fe–21Mn alloy. The asymmetric form of the peak corresponding to the γ-phase (this effect is stronger for Fe–16Mn alloy) indicates some inhomogeneity in the distribution of the atoms in the lattice of the alloy [5, 139].

Examination of the Fe–20Mn alloys, alloys with titanium and silicon, shows an important regularity. It is well-known that titanium is a stabiliser of the γ-phase in these alloys, and silicon, at a content of 2%, greatly reduces the stability of austenite in relation to the γ→ε transformation in cooling (cobalt has the same effect). Analysis of the absorption spectra of ternary alloys (Fig. 5.12) indicates that the inhomogeneity of distribution of H_{ef} in the alloys with titanium is minimum; however, in the alloy with an addition of silicon, the austenite peak has two distinctive centres of distribution of H_{ef}

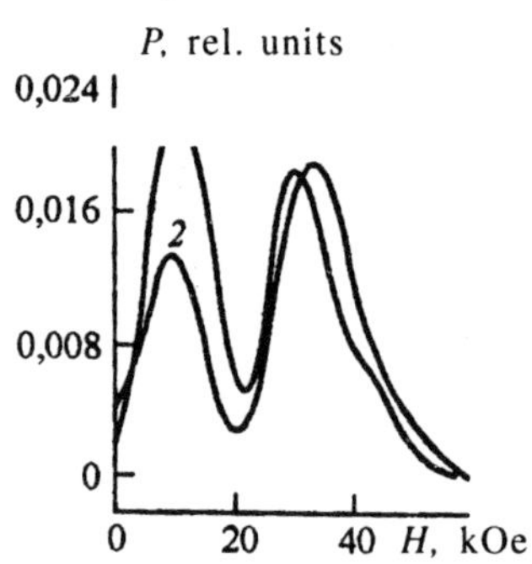

Fig. 5.11 Distribution of hyperfine fields in alloys Fe–16Mn (1) and Fe–21Mn (2).

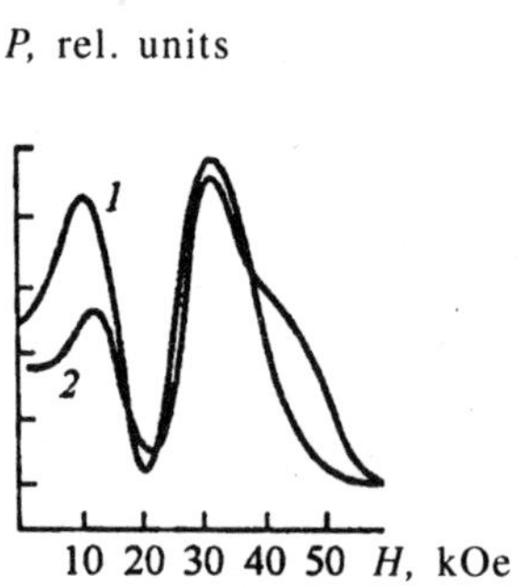

Fig. 5.12 (right) Distribution of hyperfine fields in alloys Fe–20Mn–Ti (1) and Fe–20Mn–3S (2).

which unambiguously indicates the inhomogeneous distribution of the atoms of the main components of the alloy when adding silicon, i.e. the short-range intra-phase separation of austenite. The effect of the alloying element on the nature of distribution of the atoms in the lattice of the alloy is in correlation with its effect on the stability of austenite. For example, the stabiliser of the γ-phase of cobalt influences the type of functions $P(H)$ similar to silicon [5, 71].

In the last decades, the Mössbauer effect has become one of the main methods of examining the short-range order and local atomic rearrangements in a number of systems. The method is used for the examination of the short-range order not only in conventional crystalline but also in amorphous and nanocrystalline materials, when the diffraction methods of analysis, such as x-ray, electron and neutron diffraction, are not effective.

Multiple investigations, carried out using the Mössbauer effect, have confirmed that a large number of solid solutions are not ideal, i.e. there is a deviation of the distribution of atoms in them from the statistical uniform distribution [140–165] (see also the review in [166]). This relates to both crystalline and amorphous alloys. The latter may be characterised by both the topological and composition short-range order.

For example, investigations of the solid solutions of Cr and Co in iron [140–143], etc. have confirmed the tendency towards the formation in annealing of a short-range order, i.e. short-range phase separation in respect of chromium and short-range ordering in respect of cobalt. The existence of a chemical or composition short-range order has also been detected in the amorphous alloys $(FeCo)_{75}SiB$ [144] and $Fe_{85-x}Co_xB_{15}$ [145].

In [140], it was found out that during low-temperature annealing at the initial stage of phase separation of the Fe–15Cr–Co–Al–Nb–V–0.25C alloy with a higher chromium content a strongly magnetic α-phase is formed, and the composition of the phase was determined. The authors of [141], using Mössbauer spectroscopy and other methods, have established that the ageing of chromium-containing Fe–15Cr–5Ni–2Cu–Ti–0.08C steel in the temperature range 375÷475°C results in the phase separation of the α-matrix into two solid solutions with different chromium content, and this is accompanied by the stabilisation of retained austenite (up to 50÷70% at the temperature of maximum hardening).

On the basis of modelling the form of the curves of resonance absorption [142] it has been shown that it is possible to separate, by means of the Mössbauer effect, the processes of short-range breakdown taking place by means of nucleation and growth, and the processes of spinodal type. The spinodal mechanism of the thermal breakdown of the $Fe_{50}Cu_{50}$ (at%) metastable alloy, prepared by mechanical alloying by vibration milling for 400 hours, by the method of nuclear gamma resonance, was confirmed in [143] Various stages of the breakdown process have been studied.

The authors of [146] used the Mössbauer effect for the examination of metastable alloys $Ti_{1-x}Fe_x$ ($0.01 \leq x \leq 0.02$), quenched from different temperatures (650÷1000°C). It has been shown that, in addition to the β-phase, a certain concentration range may also contain regions of ω- and α-phases with the short- and long-range atomic orders.

In [147–149], the examination by the method of nuclear gamma resonance and field ion microscopy established that the process of austenitization of chromium steels (Cr > 3%, C ~ 0.3÷0.4%) results in the phase separation of austenite into zones with low and high content of chromium. In quenching, the inhomogeneity of austenite is inherited by martensite. This fact is very interesting and important because it was assumed that the increase of the austenitization time (at a relatively high temperature) always increases the degree of homogeneity of the solid solution (i.e. homogenising takes place).

Investigations carried out in [167] revealed the phenomenon of hysteresis on the curves of temperature scanning of the number of the recorded γ-quanta in the range of the paramagnetic–antiferromagnetic phase transition in the Fe–18Cr–18Mn–0.4N–0.1C and Fe–18Cr–0.4N–18Mn–5Ni–Mo–V–0.1C austenitic steels. The width of the hysteresis loop depended on the degree of alloying and additional thermal mechanical treatment. Mössbauer examination, x-ray diffraction analysis and mechanical tests showed the instability of the properties of the steels in respect of time during holding in the antiferromagnetic state. The results of analysis of the data indicate short-range phase separation of austenite into zones with different content of chromium, which also cast doubts on the conventional considerations treating austenite as a homogeneous solid solution.

Very important data have been obtained in the Mössbauer examination of radiation-stimulated processes of the redistribution of atoms in corpuscular radiation (see also chapter 6).

For example, in [150–152] the authors indicated several unique possibilities of Mössbauer spectroscopy for the examination of radiation damage in metals and alloys under corpuscular irradiation and the associated processes of redistribution of the atoms. In particular, the specific role of the zone of stopping of the primary knocked out atoms has been examined. Similar data could not be obtained by other methods. It has been shown [150, 151] that the irradiation of Ni_3Fe intermetallic compound with charged particles results in the formation of precipitates of Ni, Fe. To explain the phenomenon, the authors used the model of radiation separation, and evaluated the critical radius of stabilisation of precipitates (~26 Å). Identical disruptions of the stoichiometry of the alloys were detected for Fe_3Al and Ni_3Mn. A strong dependence of the nature of the effect of radiation on the amorphous alloys on their composition has been established [152]..

The authors of [153, 154] investigated radiation-stimulated processes in Fe–Ni alloys of the invar composition, with additional alloying by other elements. It has been shown [153] that ~2.6 wt.% of Ti changes the type of radiation-stimulated transition in low temperature (393÷473 K) irradiation with electrons (5.5 MeV). The phase separation of the FCC solid solution in respect of nickel is replaced by the precipitation of the high-nickel N_3Ti intermetallic phase. Identical results were obtained for an Fe–Ni alloy, additionally alloyed with aluminium (precipitation of Ni_3Al phase) [154]. In the same study it was established that in alloying with Si and Zr intermetallic ageing is almost completely absent, and a certain amount of intermetallics is already present in the initial (quenched) state.

In [168], the method of nuclear gamma resonance was used for the analysis of the atomic structure of $Fe_{100-x}Cr_x$ alloys (x = 1.7÷48.1 at.%) after quenching and subsequent annealing at 670÷770 K, and also after irradiation with 5.5 MeV electroncs at 320÷570°C. The experimental results show that in the vicinity of the concentration of 10 at.% Cr, there are two types of short-range order: short-range ordering at $x < 10$ at.%, and short-range separation at $x > 10$ at.%. This corresponds to the change of the sign of the Cowley–Warren short-range order parameter in the first coordination sphere.

In [169, 170], Mössbauer spectroscopy was used for the examination of the processes of formation of new phases in iron alloys with silicon, chromium and nickel irradiated with electrons with a subthreshold energies (20÷30 keV). The authors reported that in the conditions which almost completely exclude direct defect formation, the degree of development of the processes of redistribution of the atoms is considerably higher in comparison with the case of high-energy radiation. Assumptions have been made on the existence of the role of collective effects and synergic processes in 'open systems' (subjected to the continuous external effect of radiation).

As already mentioned, the existence of the chemical (composition) short-range order was detected by the Mössbauer effect in the alloys (Fe, Co)$_{75}$SiB$_{15}$ and Fe$_{85-x}$Co$_x$B$_{15}$ [144, 145], and also in a large number of other amorphous and nanocrystalline alloys (the distribution of the atoms of titanium, vanadium, chromium, manganese, cobalt, nickel, silicon, etc.) [144, 145, 171–178] was analysed). This is the only method of analysis of the processes of redistribution of the atoms subjected to different external effects in such systems.

The Mössbauer examination [179] of the Fe$_{73.5}$Cu$_1$Nb$_3$Si$_{13.5}$B$_9$ alloy, amorphised by quenching from the melt on a spinning disc, showed that the effect of both tensile loading and of the magnetic field during nanocrystallizing annealing result in changes in the content of chemical elements in phase components of the nanocrystalline alloy (in comparison with conventional annealing, taking place in the absence of external effects). On the basis of a decrease in the width of the lines of the FeSi solid solution it has been assumed that the solid solution may be ordered, and also that its Nb and B content may decrease. The observed changes in the low-field components of the Mössbauer spectrum, corresponding to the phases with a higher content of the alloying elements, indicate the increase in these phases of the concentration of 'non-magnetic' elements and the decrease of the iron content.

Special attention should be given to the studies concerned with the application of the Mössbauer effect for the examination of the distribution of interstitial impurities in metallic solid solutions.

Gavrilyuk, Nadutov *et al* [155–157] examined in detail special features of the interaction and distribution of carbon and nitrogen in iron-based alloys. These investigations, carried out using Mössbauer spectroscopy, established a principal difference in the electronic structure of Fe–N and Fe–C alloys of the FCC and BCC (BCT) lattices, which is reflected in the differences in the hyperfine electrical and magnetic structures of Mössbauer spectra. It has been shown that the interaction between the interstitial atoms is of the long-range character. The effect of N–N repulsion in the first and second coordination spheres in the BCT structure is stronger than the C–C repulsion. The situation in the FCC solid solution is complicated by the fact that N–N repulsion is stronger in the first coordination sphere, and in the second coordination sphere the situation is reversed. The results show that the distribution of the nitrogen atoms in austenite and martensite is ordered in comparison with the sub-microinhomogeneous distribution of the carbon atoms (this is attributed to the stronger resultant repulsion between the nitrogen atoms). It has been shown that alloying elements have different effects on the interaction and distribution of the N and C atoms. The resultant data explain the higher thermal stability of nitrogen austenite. It has also been shown that the character of distribution of the interstitial atoms depends strongly on the number of defects in the crystal.

Examination by transmission electron microscopy and Mössbauer spectroscopy showed [158] that the tempered N and C+N martensites in alloys based on Fe +15% Cr +1% Mo with 0.3% C, N are characterised by the short-range ordering of the chromium atoms which forms mainly in the austenite because of the special role of nitrogen and is inherited by martensite.

The positive effect of carbon (~1.5%) on the mechanical and invar properties of the Fe+30.4% Ni alloy in [159] is also explained by the effect of carbon on the atomic and magnetic ordering in this alloy.

It was reported in [160] that in the Fe +13% Mn, 0.5% C steel, plastic deformation changes the distribution of the effective magnetic field $P(H)$ on the nuclei of the resonant atoms ^{57}Fe. This is caused by the redistribution of the C atoms from the octahedral interstitials, surrounded by the Fe atoms, to the octahedral interstitials, surrounded by the Mn atoms.

The Mössbauer effect was used for investigating the distribution of C and N in the austenite of steels containing Ni and Mn [161]. The results show that the number of 90° configurations of the C–C atoms in the neighbouring adjacent interstitials is considerably lower than in the case of the random distribution of carbon. This is associated with the strong repulsion of the C atoms. The introduction of Ni increases the number of the C–C couples because nickel repulses the carbon atoms. The ordered structure of the type $(Fe, Ni)_4C$ has not been detected. The Fe–Mn–C alloys are characterised by the formation of clusters, rich in the atoms of Mn and C. They are also characterised by the presence of single carbon atoms in the vicinity of the iron atoms in the intercluster region. The distribution of the nitrogen atoms in austenite is similar to the ordered structure of Fe_4N with vacant interstitials because of a shortage of N atoms. The nature of increased stability of nitrogen austenite in comparison with carbon austenite is attributed to the more uniform distribution of N in the γ-solid solution.

Litvinov *et al* [180–182] used the Mössbauer effect for the examination of the distribution of carbon atoms in steels of different type in connection with concrete features of their physical and mechanical properties. Investigations were carried out on Hadfield steel (Fe+13.24% Mn, 1.16% C) in the state after quenching from 1050 °C [180]. The results show that at room temperature the carbon atoms are distributed in octahedral voids of the lattice and this distribution differs from the statistically uniform distribution. Examination showed changes in the hyperfine structure of the Mössbauer spectra of Fe–0.7C steel (Fe + ~7% C) as a result of plastic deformation [181] which is explained by the fact that the carbon atoms become 'invisible' at a high density of defects forming clusters on the dislocations. In [182] the method of conversion Mössbauer spectroscopy was used to examine the effect of abrasive treatment and low-temperature tempering on the distribution of carbon atoms, the stability and hardening of manganese and chromium austenite. The experimental results show that the short-range phase of separation of austenite in respect of carbon, characteristic of quenched chromium–manganese steels, becomes more intensive in low-temperature tempering and under the abrasive treatment in accordance with the Brinell–Haworth method because of heating. This influences the deformation stability and hardening of austenite.

The author of [162] analysed the Mössbauer spectra of carbon iron–manganese austenite taking into account the difference in the values of the tensor of the gradient of the electrical field for different configurations of the iron atoms surrounded by the implanted atoms. Examination of the quenched steel showed the short-range ordering of carbon atoms. Deformation of the quenched steel by filing disrupts the regions of the short-range order in respect of the angle, i.e. the steel is hardened by the well-known Fisher mechanism. It has been shown that deformation by filing initiates the ageing of steel at room temperature – after 10 years, the structure of the spectrum contains a ferromagnetic phase. It has been shown that manganese also undergoes short-range ordering whose type and degree may be determined by experiments.

The Mössbauer effect has been used [163] for the investigation of changes in the structure of cementite taking place during annealing in the α-state of 'fresh' pearlite of carbon steel, produced at the lowest possible temperature of pearlitic transformation and the shortest holding time which is however sufficient for completing the transformation. It has been shown that these structural changes are accompanied by large changes in the Mössbauer spectra, indicating the redistribution of carbon in cementite. This means that the cementite, formed at relatively low temperatures of pearlitic transformation, has a structure differing from that of stable cementite. The changes in the structure of cementite are accompanied by changes in the ferritic component of pearlite and the mechanical properties of steel.

The methods of Mössbauer spectroscopy, x-ray diffraction analysis and transmission electron microscopy [164] have been used to examine the distribution of nitrogen atoms in the process of ageing and tempering of quenched Fe–N steels with a high nitrogen content. It has been shown that the distribution of the nitrogen atoms depends on the method of nitriding (gas nitriding or ion implantation) and also that the distribution of the atoms in the surface layer whose thickness was smaller than 200 nm differs from the distribution in the volume. The main stages of ageing and tempering have been determined: ordering of the nitrogen atoms with the formation of small coherent precipitates α''–$Fe_{16}N_2$; transition, as a result of thickening, to semi-coherent particles α''-$Fe_{16}N_2$; the solution of α''-$Fe_{12}N_2$ with the formation of γ-Fe_4N; breakdown of retained austenite during tempering. The first and third stages are characterised by the activation energy identical with the activation energy of diffusion of nitrogen, the second stage is characterised by the diffusion of iron on the dislocation nuclei.

In [165], concerned with the investigation of austenite, alloyed with nickel and manganese, a correlation between the short-range atomic ordering of austenite and the high tetragonality of freshly formed martensite has been revealed. The results also show large differences in the nature of redistribution of the atoms of C and N at $T > 200$ K. The Mössbauer effect testified to the formation of clusters of carbon atoms and the ordering of nitrogen atoms.

The method of nuclear gamma resonance makes it possible to examine the local atomic and magnetic structure of the grain boundaries in polycrystals [183–191] and also special features of the state of microvolumes, adjacent to these boundaries [192–197].

Kaigorodov, Klotsman *et al* [183–190] investigated the type of states populated with the atomic probes ^{57}Co in the course of intercrystalline diffusion of these probes in polycrystalline metals (3*sp*-: Al, 3*d*-: V, Cr, Cu; 4*d*-: Nb, Mo, Rh, Pd, Ag; 5*d*-: Ta, W, Ir, Pt, Au). The experimental results show that the states, localised in two regions of the zone of intercrystalline diffusion are populated: in the 'nuclei of the grain boundary' and in the near-boundary regions of the regular lattice. In the equilibrium conditions, the population of these two states is determined by the ratio of the atomic volumes of ^{57}Co and populated positions. These two types of state differ in the parameters of the electronic structure and in dynamic properties [187, 188].

In [191], approximately 1% of iron, enriched with ^{57}Fe, was introduced into W by mechanical alloying. Different components, found in the Mössbauer spectrum, were identified as the atoms of iron, distributed: 1) in the substitutional positions in the crystal lattice; 2) in the vicinity of lattice defects; 3) at the grain boundaries.

In [192–197], the Mössbauer effect was used to detect two states of the iron atoms with different parameters of the electrical and magnetic hyperfine structure in the sub-microcrystalline iron whose structure was formed as a result of compacting or shearing with subsequent annealing. One of these states corresponds to the iron atoms in conventional iron. The second state is regarded as a state corresponding to the 'grain boundary phase', characterised by special physical properties (electrical, magnetic, thermophysical, etc) and by a specific physical width.

The high sensitivity of the Mössbauer effect to the local atomic environment makes it possible to examine not only the local atomic and electronic structure of inorganic crystals but also biological objects. For example, in [198–200], using Mössbauer spectroscopy, unique data were obtained on special features of hemoglobines and iron–dextrane complexes with different molecular structure, including the hemoglobines of human beings.

5.3. EXAMINATION OF THE MAGNETIC STRUCTURE OF ALLOYS

5.3.1. Iron-based solid solutions

The parameters of the hyperfine structure of metallic solid solutions were determined on the basis of modelling of the form of the external peaks of Mössbauer spectra taking into account interactions in the radius of two [201], three [202] and six [32, 34] nearest coordination spheres of the iron atoms. This approach is restricted by the framework of the externally applied model of hyperfine interactions. In addition to this, even if the absolute adequacy of the model is not doubted, the optimality of the resultant solution for a large number of the determined parameters cannot be guaranteed because of the cumbersome problem of global examination of the target function and determination of its global maximum (Appendix 1).

Because of the above, it is interesting to examine the results of restoration of the visually undetectable components of the magnetic hyperfine structure of solid solutions based on iron [139, 203, 204] with the application of the method of model-free processing of the curves of resonance absorption using the minimum of a priori information and described in Appendices 1 and 2. In [203, 204], these methods were used to determine the parameters of hyperfine interaction for the atoms of the first and indirectly (on the basis of the increase of the width of the line) second coordination spheres of ^{57}Fe (in the alloys of the Fe–Al system), and in [79] – the first, second and indirectly third coordination spheres (in the alloys of the Fe–Cr system).

The authors of [203, 204] analyzed the magnetic hyperfine structure of Mössbauer spectra of iron–aluminium alloys containing 5, 10, 15 and 20 at.% aluminium. The functions of probability density $P(x)$, reflecting the distribution of the centres of gravity of Lorentzians, forming the external peaks of the Mössbauer spectra, were restored. In terms of the theory of solution of inverse problems of spectroscopy, it was possible to restore 'true spectra' free from instrumental and physical broadening of the spectral lines. This procedure makes it possible to clarify the regularity of the effect of different configurations of the nearest environment of the resonance nucleus on the value of H_{ef} without the application of modelling considerations regarding the nature of hyperfine interactions and the fine structure of the spectrum. In addition to this, it is also possible to determine the probabilities of the ^{57}Fe nuclei surrounded by different numbers of atoms of Fe and Al. The functions $P(x)$ were restored using the procedure described in Appendix 1. The optimum number of the terms of expansion of the Fourier series was 15–17 (see the relationship (A1.16)).

The external peaks of the Mössbauer spectra and the functions $P(x)$, determined from the spectra, are presented in Fig. 5.13. Their shape indicates the presence of several centres of distribution of the parameters of hyperfine interaction. The values of the x-coordinates of the maxima of the functions $P(x)$ are identified as the external peaks of the Zeeman sextets of the lines, corresponding to the configurations of local environment with l =0,

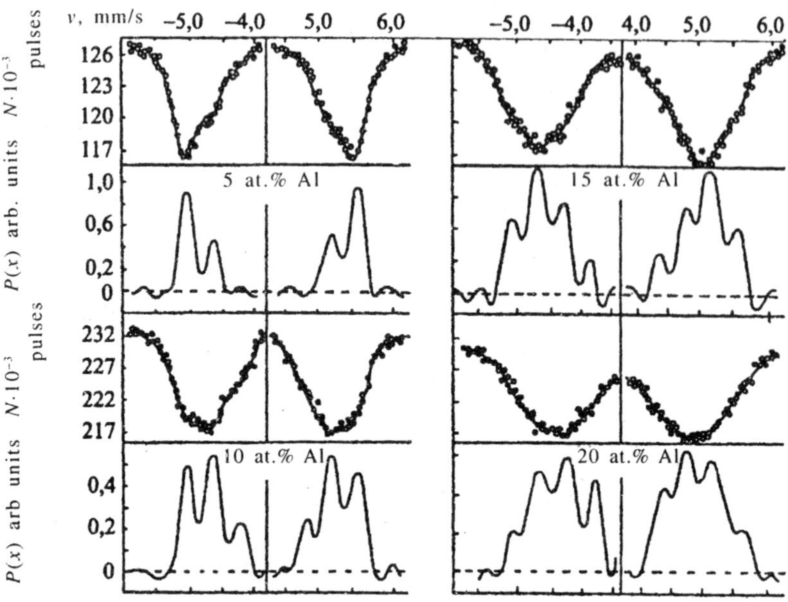

Fig. 5.13 External peaks of Mössbauer spectra and the densities of distribution of elementary components calculated from these peaks.

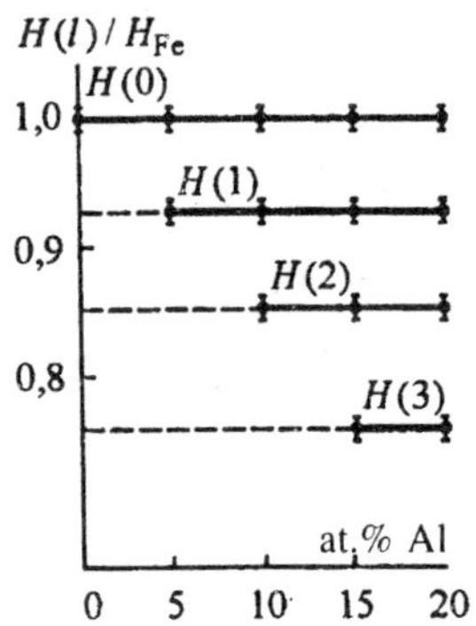

Fig. 5.14 Concentration dependence of the strength of the effective magnetic field on Fe nuclei with different numbers of the nearest aluminium atoms (H_{Fe} = 331 kOe).

1, 2, 3 atoms of aluminium in the first coordination sphere of iron. The measured values of $H(l)$ for the alloys of different composition are shown in Fig. 5.14. The results are not associated with the application of a priori considerations regarding the atomic–crystalline structure and special features of hyperfine interactions in the investigated system. It is interesting to note the fact that in contrast to the data in [205], there is no significant dependence of the values of $H(l)$ on the aluminium concentration up to 20 at.% aluminium (Fig. 5.14). Evidently, this indicates the mutual compensation of alternating contributions from the second and more remote coordination spheres (paragraph 3.2, 3.3). At the same time, the experimental results confirm the presence of a small non-linearity in the dependence of H_{ef} on the number of aluminium atoms in the nearest environment. The measured contribution of the nearest aluminium atom to the isomeric shift is ~+0.02 mm/s.

Comparison of the areas below the well-resolved peaks $P(x)$ with the results of calculation of the probabilities of the surrounding assuming the random distribution of the atoms of iron and aluminium indicates the strong tendency to the development of short-range ordering which is most distinctive at a higher aluminium content. In [204], the authors calculated the values of the short-range Cowley parameter α_1 in the first coordination sphere for Fe–Al alloys of different composition annealed for 3 h at 500 °C, on the basis of the type of $P(x)$ functions and also on the basis of the modelling approximation of the spectra using the data of model-free processing as the initial approximation. The experimental results show that the shape of $P(x)$ curves strongly depends on the prior thermal history of the specimens. For example, the specimens, subjected to holding at 800°C (followed by quenching), are characterised by the presence of well-resolved, symmetric peaks $P(x)$. Annealing at 500 and 600°C resulted in the distortion of the form of the peaks (in a number of cases it resulted in distinctive asymmetry). The latter is in agreement with the data published in [86, 206, 207] according to which the alloys of the Fe–Al system, containing more than 10÷15 at.% aluminium, may be characterised by the occurrence of processes, resulting in the formation of zones with different concentration of aluminium and the type of order.

The authors of [79] analysed local disruptions in the magnetic structure and the distribution of atoms in Fe–Cr, Fe–Co and Fe–Cr–Co alloys by the method of restoration of the density functions $P(x)$. The application of the methods of model-free analysis of the alloys enabled the evaluation of the hyperfine structure parameters and specifying the role of the first and more remote spheres in the examined pattern of hyperfine interactions. For the first time, the type of density functions $P(x)$ was used for visual 'resolution' of the effect of the second coordination sphere. Analysis of the density functions confirms conclusions on the formation of the short-range atomic order, made on the basis of the result of modelling treatment of gamma resonance spectra [59].

The external peaks $+1/2 \rightarrow +3/2$ and $-1/2 \rightarrow -3/2$ of the Mössbauer spectra of the iron alloys with 13.6 at.% chromium after two different heat treatments: a) quenching from 1000°C; b) quenching from 1000°C, ageing at 450°C, 100 hours, and the functions of the density of distribution of the centres of gravity of the elementary components, obtained from these spectra in [79], are presented in Figs. 5.15 and 5.16.

The calculated curves $P(x)$ demonstrate the very high degree of resolution. Figure 5.15 shows the results of identification of the lines of the hyperfine structure of the external peaks of the

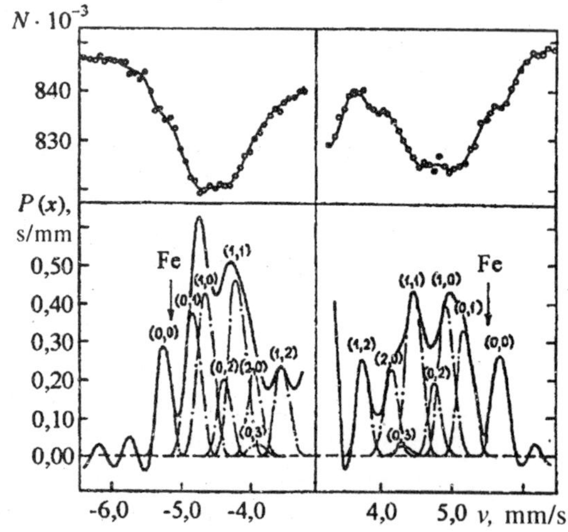

Fig. 5.15 External peaks of the Mössbauer spectrum of quenched Fe–13Cr alloy and the functions $P(x)$ of the density of distribution of elementary components restored from the spectra; (l_1, l_2) denote the number of Cr atoms in the first and second coordination spheres of iron atoms.

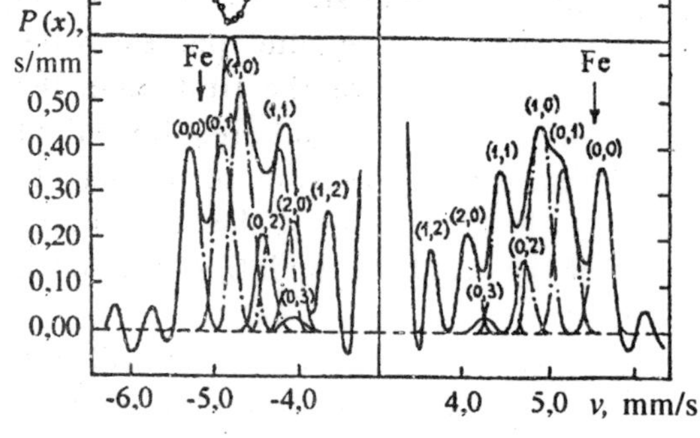

Fig. 5.16 External peaks of the Mössbauer spectrum of the Fe–13Cr alloy annealed at 450 °C for 100 h and the functions $P(x)$ of density of distribution of elementary components restored from the spectra.

Mössbauer spectrum of the quenched alloy on the basis of the form of the restored density functions. The indexes in the round brackets indicate the number of the chromium atoms in the first l_1 and in the second l_2 coordination spheres of the iron atom. The values of the effective magnetic fields, isomeric shifts and estimates of the probability $P(l_1, l_2)$, obtained on the basis of the analysis of the type of functions $P(x)$, are presented in Tables 5.1–5.3. For comparison, Tables 5.2 and 5.3 also present theoretical values of these probabilities $P^*(l_1, l_2)$, calculated assuming the absence and presence of the short-range atomic order. As indicated by the comparison of these data, the distribution of the atoms in the quenched alloy is close to the statistically uniform distribution, which is in agreement with the results of model-based processing of the spectra [59].

Figure 5.16 shows the results of 'model-free' analysis of the external peaks of the spectrum of an iron–chromium alloy with 13.6 at.% Cr, aged at 450°C for 100 h. As in the case of the quenched alloy, the type of produced functions $P(x)$ makes it possible to reliably separate the effect of the first two coordination spheres (the nearest and next nearest neighbours of the resonant atom of ^{57}Fe). The results of analysis of the hyperfine structure and of the determination of the probability values are presented in Tables 5.1 and 5.3. The changes in the probabilities in the process of ageing of the alloy confirm on the whole the results of model-based processing of the spectra in [59]. The best description of the experimentally determined values of the probabilities by the product

Table 5.1 Values of the parameters of the hyperfine structure of the Fe+13.6 at.% Cr alloy determined on the basis of analysis of the type of functions $P(x)$

(l_1, l_2)	Effective magnetic field, kOe		Isomeric shift in relation to α–Fe, mm/s	
	Heat treatment			
	Quenching from 1000 °C	Annealing at 450 °C, 100 h	Quenching from 1000 °C	Annealing at 450 °C, 100 h
(0.0)	341_5	341_5	0.03_2	0.03_2
(0.1)	310_5	313_5	0.01_2	0.01_2
(1.0)	299_5	299_5	-0.05_2	-0.05_2
(0.2)	278_5	279_5	-0.01_2	-0.01_2
(1.1)	271_5	272_5	-0.08_2	-0.04_2
(2.0)	253_5	254_5	-0.13_2	-0.13_2
(1.2)	240_5	228_5	-0.15_2	-0.13_2

Table 5.2 Values of the probabilities of the atoms of Fe being surrounded by atoms of Cr for the quenched alloy Fe + 13.% at.% Cr determined on the basis of analysis of the type of functions $P(x)$ and calculated for $\alpha_1 = \alpha_2 = 0$

$P(l_1, l_2)$ (calculated for $P(x)$)				
l_1/l_2	0	1	2	3
0	0.083	0.117	0.054	0.028
1	0.136	0.152	0.077	–
2	0.067	–	–	–
3	–	–	–	–
P^* (l_1, l_2) (calculated for $\alpha_1 = \alpha_2 = 0$)				
l_1/l_2	0	1	2	3
0	0.088	0.100	0.047	0.012
1	0.134	0.152	0.072	0.018
2	0.089	0.101	0.048	0.012
3	0.034	0.038	0.018	0.005

Table 5.3 Values of the probabilities of atoms of Fe being surrounded by Cr atoms for the Fe + 13.6 at.% Cr alloy, annealed at 450 °C, 100 h, obtained on the basis of analysis of the type of functions $P(x)$ and calculated for $\alpha_1 = \alpha_2 = 0.20$

$P(l_1, l_2)$ (calculated for $P(x)$)				
l_1/l_2	0	1	2	3
0	0.148	0.164	0.064	0.034
1	0.197	0.152	0.065	–
2	0.079	–	–	–
3	–	–	–	–
P^* (l_1, l_2) (calculated for $\alpha_1 = \alpha_2 = 0.20$)				
l_1/l_2	0	1	2	3
0	0.148	0.130	0.048	0.009
1	0.173	0.152	0.056	0.011
2	0.089	0.78	0.029	0.005
3	0.026	0.023	0.008	0.002

of the binomial distributions (see the relationship (4.41)) is obtained at $\alpha_1 = \alpha_2 = 0.20$.

Analysis of the resultant functions of density for the quenched and aged alloys indicates that the model of the additive effect of the chromium atoms on the value of H_{ef} is fulfilled with acceptable accuracy. The position of the lines of the hyperfine structure is

Table 5.4 Values of the addiditive contributions to H_{ef} on the ^{57}Fe nucleus from Cr atoms calculated from the type of functions $P(x)$

$h_1 = \Delta H_1/H_{Fe}$	$h_2 = \Delta H_2/H_{Fe}$	$h^* = [H(0,0) - H_{Fe}]/H_{Fe}$, kOe
-0.133_5	-0.093_5	$\sim +0.033$

practically independent of the type of heat treatment.

Table 5.4 shows the values of the additive contributions h_1 and h_2 to the effective magnetic field on the ^{57}Fe nucleus. Comparison of these data with the data obtained in [35, 59] on the basis of modelling and fitting of the form of the Mössbauer spectrum indicates that they greatly differ. This is natural because the search for the minimum of the non-linear functional, carried out in the process of model-based processing, is the 'non-trivial problem' [208], and there is a probability of hitting one of the secondary minima of the target function. The exact solution of the problem is guaranteed only by specifying the sufficiently good initial approximation. The results of application of the method of restoration of functions $P(x)$, not associated with the application of physical models and not requiring initial approximation, should obviously be regarded as indication of the need for improving the accuracy of the values of the parameters of the hyperfine interaction for the Fe +13.6 at.% Cr alloy. It should be remembered that the values of the contributions h_1 may greatly depend on the concentration (see chapter 3).

The positive sign of h^* in Table 5.4 is in agreement with the predictions of the Ruderman–Kittel–Kasui–Yosida theory (see [21, 36, 47]). On the whole, as shown in [21], the highest in absolute value contribution to H_{ef} in the group of the rejected contributions ($n \geq 3$) from the third coordination sphere is positive. In addition to this, a significant positive contribution to $H(0, 0, 0)$ may also be provided by the effects of magnetisation (repopulation) of conduction electrons (chapter 3).

Similar results were obtained in [209] for the Fe +13 at.% Cr +14 at.% Co ternary alloy and Fe +14 at.% Co binary alloy. The Fe–Co alloy is characterised by the single peak $P(x)$. The latter is associated with the fact that the magnetic state of cobalt in this alloy is very close to the magnetic state of Fe ($\mu_{Co} \approx \mu_{Fe}$, see Table 3.2, paragraph 3.3.4) and, in addition to this, the perturbations of the magnetic moments of the adjacent iron atoms caused by cobalt are small.

The studies carried out mainly in the first two decades after the discovery, by R. Mössbauer, of the phenomenon of nuclear gamma resonance, formed the basis of the Mössbauer methods of analysis of the local atomic and magnetic environment of the nuclei of the resonance isotope and also of the short- and long-range atomic and magnetic order in the alloys.

Up to now, a large number of studies have been published which confirmed all the main experimental results and theoretical representations, described in chapter 3 and also in the present chapter, concerned with the nature of the effective magnetic fields on the nuclei of the atoms and the magnetic hyperfine structure of nuclear gamma-resonance spectra.

Investigations of the properties of Fe_{1-x}(Cr, V) alloys [210], carried out using the Mössbauer effect have shown that, as in the case of binary Fe–V, Fe–Cr alloys, special features of the variation of their magnetic properties may be explained on the basis of the model which describes the oscillations of the spin density of the conduction electrons determined by changes in the magnetic moment on the impurity node.

In [211], the spectra of $Au_{1-x}Fe_x$ (x = 0.14 and 0.18) in the temperature range 5÷300 K and external fields of 0.2÷0.4 T (2÷40 kOe) were analysed taking into account the anisotropy of the magnetic hyperfine interaction in the atomic configurations with a non-spherical distribution of electronic density. The proposed model was used for the theoretical calculations of the distribution function of magnetic hyperfine and exchange fields, acting on the iron atom. Examination showed a marked spatial inhomogeneity of the exchange interaction, being the result of the strong indirect exchange Fe–Fe at small distances and of a stronger alternating interaction with a large radius. The results supplement the existing data.

The authors of [212] observed large differences in the concentration dependence of the effective magnetic fields and of the mean magnetic moment for supersaturated BCC FeSn alloys from the similar dependence for FeSi. Examination showed a weak dependence of the mean magnetic moment (calculated per single atom) on the concentration of Sn up to c = 25 at.%. According to the results presented in chapter 3 (Table 3.2), this fact indicates the presence in elements such as Ga and Sn of intrinsic magnetic moments, and also of positive exchange interactions Ga–Fe and Sn–Fe.

5.3.2. Concentrated alloys. Ferromagnetics, antiferromagnetics, non-collinear spin structures

The Mössbauer effect may be used efficiently for the examination of the magnetic properties and local magnetic structure of not only diluted but also concentrated solid solutions. For example, in [213], using the available neutron diffraction data and also the results obtained in Mössbauer experiments, it was found possible to describe satisfactorily the behaviour of the effective magnetic field in the $(Fe_{1-x}Co_x)_3Si$ ternary alloys and to select one of the three alternative models of the concentration behaviour of the nodal magnetic moments.

The methods of Mössbauer spectroscopy and x-ray diffraction were used to examine the atomic and magnetic structure of the $Mn_xFe_{1-x}Pd$ ($0< x< 1$) ternary alloys [214]. The results show that the state of the alloys is of the two-phase type (each phase has the tetragonal structure of the $L1_0$ type). Using the magnetic methods, it was shown that these alloys are characterised by the presence of three types of magnetically ordered regions: with the ferromagnetic structure of the FePd type, the antiferromagnetic structure of the MnPd type, and the 'intermediate' type of structure formed in the vicinity of the interface between the ferromagnetic and antiferromagnetic phases.

The magnetic properties of concentrated alloys FeCu, FeAl and YFe_2, produced by mechanical alloying, were investigated in [215–219]. In [220] comparative Mössbauer examination of the magnetic state of ultrafine particles and massive FePt alloy was carried out.

The method of nuclear gamma resonance [221] was used to study the ordered FePd and FePt alloys of the equiatomic composition. It was shown that the special features of the spectra of these alloys are determined by the presence in them of ordered twins and recrystallised regions. Taking into account the role of exchange energy in the system of disperse twins, it was established that the direction of the vector of magnetisation in the general case should not coincide with the axis of easy magnetisation. The mean angle of deflection of the magnetisation vector from the axis of easy magnetisation was determined by experiments. The experimental value is in good agreement with the calculated data.

Tsurin *et al* [222–228] used the Mössbauer effects for the study of the magnetic properties of alloys of the Fe–Ni–Mn and Pd–Fe systems, including the state of their saturation with hydrogen. Examination showed a substantial effect of hydrogenation on the magnetic structure and properties (magnetisation and Curie temperature) of these alloys. $Fe_{70}Ni_{30-x}Mn_x$ alloys showed [222, 223] the formation, in addition to the ferromagnetic phase, of phases with a different type of magnetic order (para- and antiferromagnetic regions, and also possibly of the states of the spin glass type). It has been shown [224–220] that the introduction of hydrogen atoms into $Pd_{1-x}Fe_x$ ($x \leq 10$ at%) alloys may result in the ferromagnetic → spin glass transition. It has been reported that the formation of impurity magnetism is determined by the indirect exchange interaction of the impurity spins through the strongly correlated electrons in the narrow 3*d* zone. The potential of this interaction consists of the exponential and cosine (RKKI type) oscillating parts. Consequently, at low temperatures, the spins of the impurity atoms are ordered either in the ferromagnetic manner or in accordance with the type of spin glass. The latter is detected at low impurity concentrations when the oscillating potential prevails over the ferromagnetic potential.

Of considerable interest are the investigations of the hyperfine interactions in the metastable ferromagnetic films of stainless steel with the BCC lattice. It is well-known that the metastable films produced from 304 stainless steel by quenching from the vapour phase, are strongly ferromagnetic materials. In [229] investigations were carried out into such films deposited on mica and Al_2O_3 by magnetron sputtering, immediately after deposition, after annealing and in the conditions with pressure of up to 36 kbar. The Mössbauer spectrum indicates the presence of a set of magnetic states with different numbers of iron atoms surrounding the absorbing atom ^{57}Fe. It has been found that each 'non-iron' atom of Cr, Ni and Mn in the first coordination sphere of ^{57}Fe reduces the strength of the field of hyperfine interaction on the average by 2.3 T, and the same atoms in the second coordination sphere by 1.2 T. The BCC–FCC transition is not initiated by a pressure of up to 36 kbar. The FCC–BCC transition in heating starts at a temperature of approximately 500°C and is completed in the vicinity of 800°C.

In [230], nuclear gamma resonance was used for the examination of the nature of the effect of paramagnetic–antiferromagnetic transition on the mechanical and physical properties of nitrogen-containing steels of the Fe–18Cr–18Mn–0.4N type.

A large number of investigations have been carried out on iron-containing alloys of rare-earth elements.

In [231, 232], for the Laves phases of the variable composition

Er $(Fe_{1-x}Mn_x)_2$ and Dy $(Fe_{1-x}Mn_x)_2$ with the structure C14 there is observed the relationship of the parameters of hyperfine interaction with the nature of the nearest environment of the Mössbauer nuclei. The results show the additivity of the contributions to the hyperfine field on the ^{57}Fe nuclei from the atoms of transition metals. The value of the variation of the field in the substitution of the Fe atom by the Mn atom was determined.

The dependence of the hyperfine magnetic field on composition in connection with the substitution of the aluminium atoms by iron atoms in the compounds $Y_2Fe_{17-x}Al_x$ (type $ThNi_{17}$ at $x \leq 4$ and ThZn at $5 \leq x \leq 7$) was investigated by the authors of [233]. The results show a reduction of magnetisation and also the non-monotonic variation of the Curie temperature (increase at $x \leq 3$ and decrease at $x > 3$) with the increase of the aluminium content. The authors of [234] investigated the effect of aluminium atoms on H_{ef} in the alloys $Tb_2Fe_{17-x}Al_x$ ($x = 0, 2, 3, 4, 5, 6, 7, 8$). The results show that at 85 K all alloys are ferromagnetic. The mean weighted hyperfine field decreases almost linearly. The decrease is 21 kOe for one substituting atom of aluminium. At $x \geq 5$, spin reorientation takes place, described by the axial model of magnetisation.

In [235], the same authors used the Mössbauer effects and other methods for the investigation of the magnetic properties of a series of alloys $Nd_2Fe_{17-x}Al_x$ ($x = 2.04, 4.01, 5.97, 7.94$ and 9.06). The results show that at low concentrations, aluminium occupies $18h$ positions, and at high concentrations, $6c$ and $18f$. It has been established that the value of the contribution ΔH to H_{ef} on the ^{57}Fe nucleus from the nearest aluminium atom decreases with increase of the concentration. This is associated with a change in the period of oscillations of the spin density of the conduction electrons [31–34] (chapter 3). The variation of the isomeric shift indicates the redistribution of the $3d$ and $4s$ electrons of the Fe atoms.

Identical investigations were carried out for the $CeFe_{17-x}Si_x$ ($x = 0.0; 0.23; 0.4; 0.6; 0.8; 1.02; 1.98$ and 3.2) [236]. The Mössbauer spectrum, as in the previous studies, was described on the basis of the binomial distribution of the probabilities of the atomic environment. It has been established that the silicon atom behaves in the manner similar to that of a magnetic hole (as reported for diluted iron alloys with silicon [22]), decreasing the effective fields on the nucleus of the iron atom.

In the compounds $ErPt_2Si_2$ (structural type $CuBe_2Ge_2$) [237] and Dy_3Ni [238], investigations carried out using the Mössbauer effect determined the magnetic moments of rare-earth elements. It has been established that the magnetic moment of Er is close to the maximum moment for Er^{3+} and is equal to 8.25 μ_B. The ion Dy^{3+} in the non-collinear antiferromagnetic has the total magnetic moment $10\mu_B$ [238].

The Mössbauer investigations combined with other methods made it possible to determine the magnetic structure of intermetallic compounds Yb_3Pd_4 [239] and compounds $DyNi_2Si_2$, $DyNi_2Ge_2$, $DyAg_2Si_2$ [240] and obtain data on the parameters of hyperfine interactions.

In the last decades, interest has increased in the chemically and topologically disordered nonequilibrium structures, because of their unusual properties.

A large number of studies in this direction has been carried out using unique possibilities of the method of nuclear gamma resonance [241–244].

The methods of x-ray diffraction and Mössbauer spectroscopy [245] were used to examine the magnetic properties and the structural features of powders based on YCo_3, Y_2Co_7, YCo_5 and Y_2Co_{17}, produced by mechanical milling. In all alloys, approximately 4% of Co was substituted by the ^{57}Fe isotope. The authors concluded that both the chemical and topological order is disrupted in the process of mechanical milling in the alloys of the YCo system. This results in the formation of the amorphous state in which the probability of the Co–Co neighbourhood increases. Correspondingly, the Curie temperature and the magnetic moment of Co increase. In long-term crushing, the process of multiple fragmentation of the particles results in the manifestation of the super-paramagnetic properties of the powders. This explains the decrease of the hyperfine field, saturation magnetisation and the apparent reduction of Curie temperature.

In [246–248], the Mössbauer effect was used for the examination of the disordered specimens of the alloys $(Fe_{0.88}Mn_{0.12})_{1-x}Al_x$ (x = 0.05, 0.08, 0.11 and 0.14) and $Fe_{89-x}Mn_{11}Al_x$ (x = 5...46) because of the large differences in the emerging magnetic structures, depending on the aluminium content and heat treatment. The latter makes it possible to vary the magnetic properties of these alloys in a wide range. Ingots were homogenised at 1000°C and quenched in water. Subsequently, sheets were cut from the alloys with the low aluminium content and polished to a thickness of 20÷400 μm. The alloys with a higher aluminium content were milled in a diamond mill. In [247], the authors proposed a model for the description of the curves of resonance absorption of the investigated materials, taking into account the interaction of the ^{57}Fe isotope in the radius of the three nearest coordination spheres. The results of the description of the experimental spectra show that the long-range atomic order is not present in all the investigated alloys after these treatments. The effect of the nearest coordination spheres on the hyperfine magnetic field (in the absolute value and the sign) indicates the RKKI-like behaviour determined by the oscillation of the spin density of the conduction electrons (chapter 3). The experimental results show that for all investigated sections of the phase diagram, in relation to the aluminium concentration, there are observed different magnetic states of the alloys: paramagnetic, ferromagnetic and also the state of spin glass at a higher aluminium content and low temperature. At some concentrations, the ternary alloy may be characterised by the co-existence of different magnetic phases [246–248], whereas x-ray diffraction does not show formation of any phases with the structure differing abruptly from the non-ordered FCC matrix (for example, phases with the long-range atomic order).

After discovering the giant magnetoresistance effect in multilayer structures, interest has increased in the study of electric transfer in artificially generated systems, including nanogranular systems. The authors of [249] carried out systematic investigations of the crystal and atomic structure and also magnetic and electric transport properties of CuCo and CuFe nanocompounds using x-ray diffraction and the Mössbauer effect. Investigations were carried out on annealed heterogeneous alloys $Cu_{1-x}Co_x$, $Cu_{1-x}Fe_x$ (x = 20 and 30 at.%, respectively), prepared by mechanical alloying.

The Mössbauer spectra provide important information on the structure of $Cu_{70}Fe_{30}$ alloys. Analysis of the distribution of the effective magnetic field $P(H)$ at 77 and 300 K shows that in addition to the α-Fe lines (330 kOe), there is also a large number of configurations of the atomic environment of ^{57}Fe with a wide distribution of the hyperfine fields from 170 to 300 kOe. One of the reasons for the formation of this distribution may be the formation of the solid solution because of the higher mutual solubility of Fe and Cu, initiated by mechanical alloying. It has also been established that the structure of the alloys is in fact a nanocomposite of the disperse particles with the size of several nanometres, enriched with Co and Fe. A large part of the volume is occupied by regions with transition concentrations. The contribution of these interfaces increases with a decrease of the particle size.

After annealing, the width of the interface abruptly decreases and the particle size increases. The contributions of the interfaces to the scattering of the conduction electrons becomes less marked.

The presence of the transition regions with different concentration of Co in CuCo and Fe in CuFe nanocompounds, produced by mechanical alloying, has been confirmed by the direct examination using the method of field ion microscopy [250–251]. Of course, the transition regions in CuFe contain a large set of the non-equivalent atoms of ^{57}Fe with different numbers of the nearest atoms of Cu, which explains the formation of the wide distribution of the hyperfine magnetic fields. With a decrease of the particle size to several tens of nanometres, the regions with the pure components disappear and the zones with maximum enrichment containing close to 50% of each component.

Using the Mössbauer effect, x-ray diffraction and magnetic methods, investigations were carried out into the inhomogeneous magnetic state of the $GdFe_2$ hydrogen-amorphised alloy [252]. The mean value of the hyperfine magnetic field increases from 22.5 T for the initial compound and to 33.4 T for the amorphous hydride. The specimens amorphised at different temperatures (in the range 200÷350°C) show the same value of H_{ef}, whereas the magnetic moment, the compensation temperature and Curie temperature decrease with a drop of of hydrogenisation temperature. For the amorphous hydride, a deviation of the behaviour of the magnetic moment of the sublattice of Gd from the Brillouin dependence was observed and anomalous curves of magnetisation in strong fields were obtained. The results were interpreted assuming the inhomogeneous structural state of the hydrides.

The results of investigations carried out using the Mössbauer effect and magnetic methods [253] into the changes of the magnetic properties of the intermetallic compound $GdFe_2$ in the course of hydrogen-induced amorphisation showed a large reduction of the energy of Gd–Fe exchange interaction and an increase of the magnetic moment of Fe from 1.6 μ_B in the initial condition to 2.2 μ_B in the amorphous hydride $GdFe_2H_x$. The results were interpreted assuming the formation of a heterophase nanostructure in which the main amorphous phase is enriched with iron, in relation to stoichiometric, and is characterised by non-constant exchange interactions between the sublattices. Investigations of the hydrogen-amorphised intermetallics RFe_2H_x(R = Y, Gd–Er) [254] show that to explain their properties, in particular, to describe the temperature dependence of magnetisation, it is necessary to assume the presence of at least two constants of exchange interaction differing in the order of magnitude. In all investigated intermetallics hydrogen-induced amorphisation results in significant intensification of the Fe–Fe exchange interactions and in a decrease of the R–Fe exchange interactions.

It should be stressed that the number of studies into different aspects of the magnetism of nanocrystalline and amorphous materials carried out using the Mössbauer effect has been continuously increasing recently [255–291].

As already noted, the Mössbauer effect, because of the high sensitivity of the resonance nuclei to the local atomic and magnetic environment, makes it possible to examine very fine changes of the spin state of the alloys, including analysis of the structure of non-collinear magnetic states.

As shown in [54, 55], there is a very high sensitivity of the mean value and dispersion of the distribution of the effective magnetic field on the ^{57}Fe nuclei in the concentrated iron alloys to the variation of their atomic and magnetic structure. Initially, this was investigated on the extensively studied alloys of the Fe–Al system with the aluminium content up to 43 at.% [54]. The results show the possibility (on the basis of analysis of the functions of distribution of the magnetic hyperfine fields $P(H)$), restored from the experimental Mössbauer spectra) of reliable identification of the type of magnetic ordering of the alloy, including antiferromagnetic, transferring to pure spin-glass at $c_{Al} \rightarrow$ 50 at.%.

The form of function $P(H)$ is modelled in this case for different types of atomic and magnetic ordering and is compared with experimental relationships.

Alloys of the system $Fe(Pd_{1-x}Au_x)_3$ (x = 0.133–0.4) are extremely attractive not only from the physical viewpoint but also from the viewpoint of utilisation of the possibilities of the Mössbauer methods of identification of different types of magnetic structures. Depending on the treatment and corresponding atomic-ordered state, greatly differing spin states form in these alloys: ferromagnetic, antiferromagnetic and a number of non-collinear states, and also mixed magnetic states. The wide spectrum of the possible magnetic states is determined by the competition of direct and indirect exchange interactions of different sign for the atoms of different sorts.

In [54, 55], the Mössbauer effect was used for detailed examination of the atomic and magnetic structure of the $Fe(Pd_{0.666}Au_{0.333})_3$ alloy. The effective magnetic field on the nuclei of the ^{57}Fe atoms with different local atomic environment was specified, as is usual for concentrated alloys, taking into account the contribution of the intrinsic magnetic atom ^{57}Fe and z-projections of the magnetic moments of the atoms of the nearest environment, using the following empirical relationship:

$$H(n) = a\mu(n) + \sum_{\alpha \to \mathrm{Fe,Pd}} \sum_{\Delta_i} b_{\alpha i} P_\alpha (n+\Delta_i) \mu_{z\alpha} (n+\Delta_i) \alpha. \tag{5.3}$$

Here a, $b_{\alpha i}$ are the partial coefficients (i is the number of the coordination sphere, $(n+\Delta_i)$ is the designation of the positions in the i-th coordination sphere of the n-th resonant atom, $P_\alpha(n+\Delta_i)$ is the projection operator, equal to 1, if the position $(n+\Delta_i)$ is occupied by the atom of type α and in all other cases it is equal to 0; $\mu_{z\alpha}(n+\Delta_i)$ is the projection of the magnetic moment of the atom of type α, located in the $(n+\Delta_i)$-th position, on the axis of local quantisation of the resonant atom. Attention was given to different (including randomly non-collinear) magnetic structures taking into account the linear dependence of the value of the magnetic moment of the Fe atoms on the number of the nearest Pd atoms.

On the basis of comparison of the calculated and experimental values of $\langle H \rangle$ and D_H it was established that the ferromagnetic state forms in the Fe $(Pd_{0.666}Au_{0.333})_3$ alloy completely disordered by quenching and cold plastic deformation. In the fully ordered alloy, the antiferromagnetic state is realised both in the sublattice of Fe and in the sublattices of Pd (the remaining atoms of Au and Pd substitute statistically uniformly the sites of two other simple cubic sublattices, 'forming' the FCC lattice of the alloy). In the intermediate stages of ordering (i.e. only over the positions of Fe) randomly non-collinear magnetic structures appear. In particular, examination showed the formation of an asperomagnetic (at 77 K) structure, coexisting in different stages of annealing with ferromagnetic and antiferromagnetic structures.

In the discussed studies, the authors propose a method of determination of the density of distribution of the magnetic moments $P(\mu)$ on the nuclei of the atoms of the Mössbauer isotope ^{57}Fe. The base for the restoration of the dependence $P(\mu)$ was represented by the following relationship of this function with the density of distribution of the effective magnetic fields:

$$P(H) = \int_0^{\mu_{max}} P(\mu) P(H/\mu)\, d\mu, \tag{5.4}$$

Here $P(H/\mu)$ is the conventional density of distribution of the effective magnetic fields on the nuclei of the atoms whose intrinsic magnetic moment is equal to μ. Since the number of local configurations of the environment is relatively large, the Gaussian approximation of this function was used. The mean value and dispersion were determined on the basis of relationship (5.3). Comparison of the results of application of this method for alloys of the FeAl system with the literature data, in particular, with the neutron diffraction data, has confirmed its validity. The application of the discussed procedure for the alloys of the $FePd_2Au$ system shows the decisive role of the palladium environment in the formation of the magnetic moment of the Fe atom.

At present, the Mössbauer method, together with the method of magnetic neutron diffraction, are used widely for identification of different non-collinear spin structures [292–303], including non-collinear spin structures of the type of spin glass in crystalline, nanocrystalline and amorphous materials.

5.3.3. Interlayer boundaries in $\{A_m/B_n\}_k$ superlattices

The high sensitivity of the Mössbauer effect to the local atomic environment and the magnetic state of the atoms makes this method an irreplaceable tool in the investigation of interlayer boundaries in the thin films and multilayer magnetic superlattices $\{A_m/B_n\}_k$. This relates to the examination of the structure of interlayer boundaries (with the resolution of different configurations of the local atomic environment of the nuclei of resonant isotopes: ^{57}Fe, ^{119}Sn, ^{151}Eu, etc in the region of the interfaces) [304–306], the processes of mixing of layers during their deposition, subsequent annealing and ion irradiation (ion mixing) [307–313], magnetic properties and orientation of the magnetic moments [314–329], including the RKKI polarisation of conduction electrons (leading to spatial oscillations of their spin density) [325–329]. The latter is taken into account when explaining the special magnetic and magnetoresistive properties of superlattices.

As an example, consider the results of Mössbauer investigations of magnetic superlattice $\{^{57}Fe_m/Cr_n\}_k$, characterised by the giant magnetoresistive effect.

In [330], the authors presented the data obtained in Mössbauer investigations of the superlattices $[Fe/Cr]_{12}$/MgO (100). The combined approach with the application of model calculations and the method of restoration of the function of the distribution density of the hyperfine fields $P(H)$ was used. This enabled the procedure of consecutive deduction of subspectra corresponding to different configurations of the environment of the ^{57}Fe resonant atom to be used. A detailed structural model of the Fe–Cr transition region from the layer of 'pure' Fe to the layer of 'pure' Cr was obtained. The deviation of the magnetic moment of the Fe atoms at the interface of the layers of Fe and Cr from the plane of the layers of the superlattice was detected. The specific magnetic structure of the interface regions with different angular orientation of the magnetic moments of the Fe atoms in relation to the plane of the layers (from 0 to 90°) is linked with the co-existence of a strong antiferromagnetic interaction of the atoms of Fe and Cr and the incommensurate wave of the spin density in the layers of Cr.

Figure 5.17 shows the Mössbauer spectra (the initial and deduction spectra) and the respecitve functions of distribution of the hyperfine fields $P(H)$. The figure shows the angle of deviation of the magnetic moments of the Fe atoms (with 1, 2, 3, etc. nearest chromium atoms) from the plane of the film. It may be seen that the magnetic moments of the peripheral atoms of iron in the interface layers (with the maximum number of the nearest atoms of chromium) are characterised by the largest angle of deviation.

In modelling of the Mössbauer spectra of superlattices $\{^{57}Fe_m/Cr_n\}_k$, the authors of [81, 82] calculated the probabilities $P(l_1, l_2)$ by the Monte Carlo method (see Table 4.1). The parameters of the hyperfine interactions, in particular, the contributions h_1 and h_2 from the chromium atoms (in the first and second coordination sphere) to the effective magnetic field of the nucleus of the Fe atom, are available only for the solid solutions of chromium in iron. Therefore, these parameters were determined (as carried out in [35] for solid solutions) through fitting the experimental curves of resonant absorption by the theoretical relationships (Fig. 5.18). The results show that the hyperfine interactions in the region of the Fe/Cr interfaces somehow differ from the hyperfine interactions of the Fe and Cr atoms in the solid solutions of chromium in iron (Fig. 5.19). According to the data obtained in [35], the contributions from the nearest and next nearest neighbours $h_1 = \Delta H_1/H_{Fe}$ and $h_2 = \Delta H_2/H_{Fe}$ to the effective magnetic field on the Fe nucleus, measured for diluted solid solutions of chromium in iron, are equal to respectively –0.083 and –0.071 ($h_2/h_1 = 0.84$). The following values were obtained for the atoms of Fe in the interface range: $h_1 = -0.069$ and $h_2 = -0.009$ ($h_2/h_1 = 0.13$). The latter indicates the reduction of the period of oscillations of the spin density of the conduction

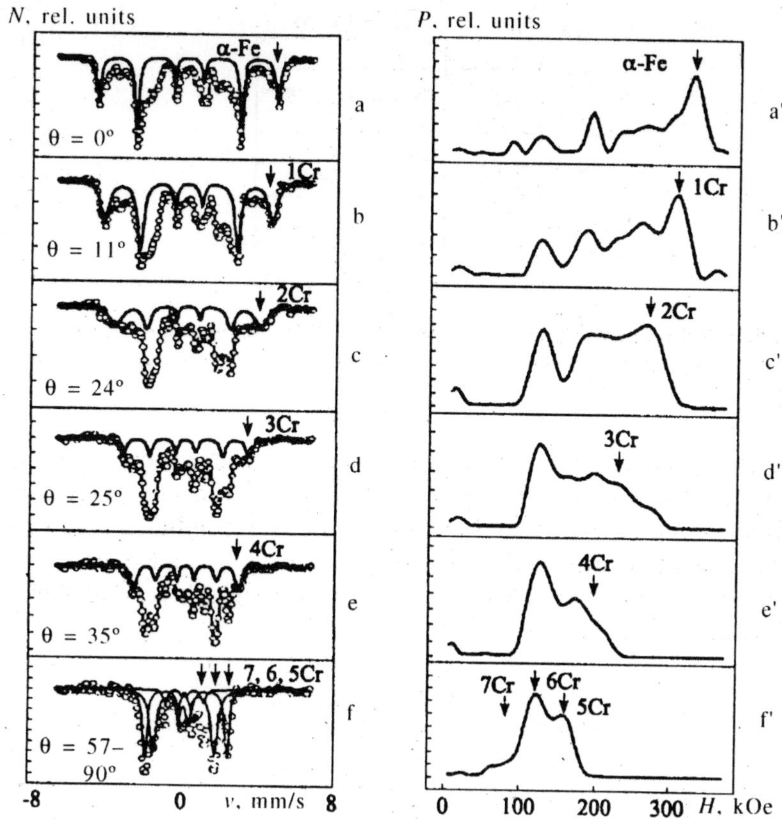

Fig. 5.17 Procedure of mathematical processing of the Mössbauer spectrum of ^{57}Fe in superlattices by the methods of consecutive deduction of subspectra S_{nCr} for different configurations n_{Cr} and the functions of distribution of the hyperfine magnetic field $P(H)$ corresponding to this procdure, using an example of superlattice $[Cr(13\ Å)/^{57}Fe\ (14\ Å)]_{12}/Cr(90\ Å)MgO(100)$.

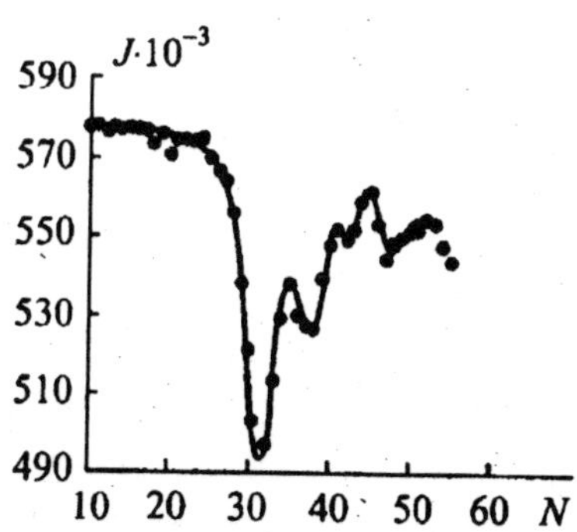

Fig. 5.18 Experimental and theoretical form of the external peak $-1/2 \rightarrow -3/2$ of the Mössbauer spectrum of the superlattice $\{Fe_8/Cr_{13}\}_k (h_2/h_1 = 0.13)$.

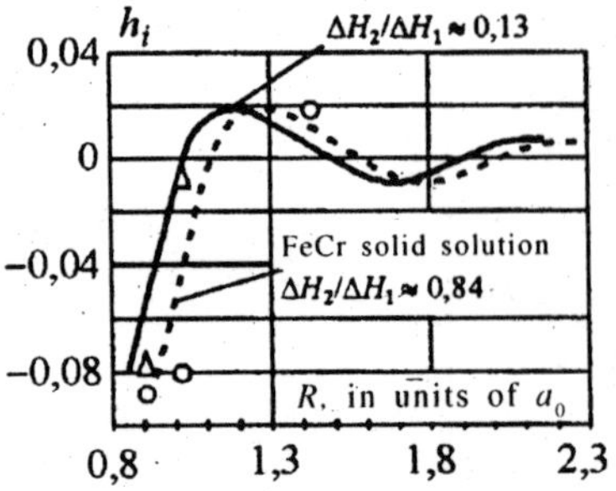

Fig. 5.19 Radial dependence of the spin density of conduction electrons for solid solutions Fe_{1-x}, $x \leq 0.25$ (dashed line) [35, 79]) and for the Fe/Cr interface (solid line).

electrons (see chapter 3) which is inversely proportional to the value of k_F of the alloy. The increase of k_F in the interface region indicates the increase of the density of the conduction electrons. This is a very interesting result. However, because of the very significant error, these data require additional verification and refinement.

Chapter 6

MÖSSBAUER SPECTROSCOPY OF ION-DOPED METALS AND ALLOYS

In this chapter, attention is given to the published (mostly in recent years) studies of the Mössbauer effect used in the investigation of the effect of beams of accelerated particles on metals, alloys, and also silicon and certain compounds. The nature and possible reasons for the detected changes in the structure and properties of the irradiated materials both in the zone of penetration of ions and in its immediate vicinity, and in subsurface layers, whose thickness is several orders of magnitude larger than the projected ion ranges, are discussed. The possibilities of the investigation of the surface and volume of solids using the Mössbauer effect are briefly analysed.

The advent of compact high-current charged beam particle accelerators [331–334], characterised by sufficiently high efficiency and capable of generation of ion beams of practically any chemical element, as well as electron beams with the kinetic energy from several tens to several hundreds of keV, has opened unique possibilities for influencing the composition, structure and physical properties of the surface of solids. The application of the methods of electron- and ion-beam treatment of materials makes it possible to modify the surface properties almost independently of the state of the volume. The latter is of special importance when we deal with the operation of solids in contact with other media, for example: solid–solid, solid–liquid, solid–gas, solid–plasma (or radiation fluxes). The physics of the effect of charged particle beams on the solids differs in principle from all the currently available physical processes. The present chapter is concerned with the review of the results of investigations of the effect of ion beams on metals, alloys and certain compounds using the Mössbauer effect.

The stopping of the ions with the above energies in solids is caused by two main processes [335–337]: elastic collisions with atoms (nuclei) of the target; the formation of cascades of atoms knocked out from their stable positions; inelastic interaction with bound electrons (excitation and ionisation effects). The contribution of other known processes is negligible. The energy transferred in this process is initially localised in the vicinity of the trajectories of the individual ions (the duration of existence of a cascade is approximately $10^{-14}\div10^{-12}$ s; the mean projected range R_p of the ions with the inidcated energies varies from several tens to several hundreds of nanometres). The energy release density may reach $10^{13}\div10^{14}$ W/cm^3.

The range of spatial and time characteristics of the processes, taking place during ion bombardment of solids, is extremely wide: correspondingly, from 10^{-14} seconds to several days on the timescale (see, for example [338]) and from the thickness of the monoatomic layer to several hundreds of micrometers over the depth of the irradiated surface layer [337, 339, 340].

Among the physical processes taking place during ion implantation, one should highlight the following [335–337, 339–345]: 1) direct sputtering (knocking out) of the atoms of the surface layer; 2) formation of primary defects in collision processes; 3) formation, during evolution of the cascades of atomic collisions, of ionisation spikes, spikes of atomic displacements, and also highly heated areas (or thermal spikes) with subsequent quenching at a rate of $10^{13}\div10^{15}$ K/s and, in some cases, with additional thermal sputtering of the surface (for these aspects, see the review in [335]); 4) alloying of the surface layer (with the transition of implanted and matrix atoms in both in the substitutional and interstitial positions), including to the concentrations greatly exceeding equilibrium concentration; 5) formation of different complexes of impurity and matrix atoms and defects; 6) vacancy swelling and void formation (blistering); 7) occurrence of radiation-stimulated diffusion, changes in the composition of the surface layer of the target (including in respect of the main components), formation of clusters, pre-precipitates, phases; 8) dissolution of the phases; 9) formation of high surface static stresses caused by the implantation of impurities and capable of initiating phase transformations; 10) general heating of the surface layers of materials (thermal phase and intraphase processes in the solid state, melting and evaporation of the surface layer, the absence of alloying effect at j >200 A/cm^2 [337]); 11) thermoelastic waves under powerful pulsed effects; 12) formation of post-cascade lattice shockwaves from individual atomic cascades in the final stage of development [341–343]; 13) excitation of the electronic subsystem of metals [344, 345] (especially important for high-energy implantation). Obviously, this list does not include all physical processes, associated with ion bombardment. Their classification is an extremely complex problem, especially, if the superposition and mutual effect of the individual processes are taken into account.

Many scientific and applied aspects of the surface modification of metals with ion beams are at the present time still in the initial stage of investigation, regardless of general increased interest and a large number of studies in this field which have appeared in recent years, and also the individual very important applied achievements. Therefore, any possibility of the direct experimental investigation of the structure of implanted layers of metals and alloys on the atomic and electronic level is very important for the development of fundamental representations regarding the physics of the effect of ion beams on materials.

One of the most efficient methods of investigation of the atomic and electronic structure of the surface and the volume of the solids is the method of nuclear gamma resonance (Mössbauer effect) [8, 12, 346–350].

6.1. MÖSSBAUER EFFECT AS THE METHOD OF INVESTIGATION OF ION-DOPED MATERIALS

The high resolution of the nuclear gamma resonance method (NGR) makes it possible to register very small changes in the energy of nuclear transitions (less than 10^{-9} eV), determined by the redistribution of electronic density in the vicinity of the resonant nucleus. Consequently, on the basis of the variation of the energy of nuclear levels, the Mössbauer effect may be used to examine the local (at a distance of several coordination spheres closest to the resonant nucleus) electronic and atomic, including 'defect', structure of materials.

The transition of a Mössbauer nucleus from the excited to ground state may take place by means of both radiation of a γ-quantum and the emission of a conversion electron. The fraction of these transitions is determined by the internal conversion coefficient. The 'shaking' of the electron shell may also result in the formation of Auger electrons, x-ray and even light quanta so that it is possible to carry out selective in-depth analysis of the surface layers of the materials [8, 346].[1] In the published studies devoted to the investigation of implanted materials using the Mössbauer effect, these procedural possibilities are still far from being utilised completely.

The Mössbauer effect is studied using emission and adsorption techniques. In the first case, the source of information on the state of solids are the Mössbauer nuclei, emitting γ-quanta, formed as a result of radioactive decay of mother nuclei (for example, the ^{57}Co isotope as a result of the electron capture and emission of a neutrino changes to the ^{57}Fe isotope in the excited state). In this case, the absorber remains the same in all experiments, and the state of the source is studied after one or the other prior process.

In adsorption Mössbauer spectroscopy, the source of information on the investigated object (absorber) are non-radioactive absorbing Mössbauer nuclei, present in the object in the natural form or introduced in the form of additions. The adsorption experiments are carried out using two geometries: transmission – with the registration of Mössbauer γ-quanta, and scattering with registration

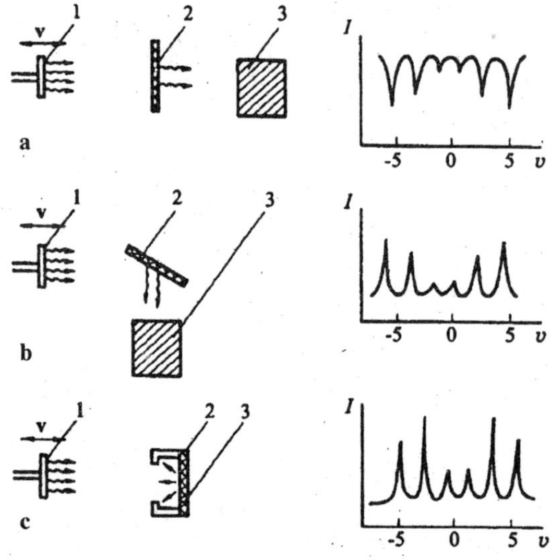

Fig. 6.1 Diagrams of Mössbauer experiments for a single (non-split) line of the source and the absorber made from a magnetically ordered material [8]: *a*) transmission geometry; *b*) scattering geometry; *c*) back scattering geometry with registration of x-rays or electrons; 1) a source moving at Doppler velocity v, 2) specimen; 3) detector. The type of magnetically split spectra is given on the right.

[1]Analysis can be made of layers situated at the depth of several tens and hundreds of nanometres (in registration of Auger and conversion electrons) to several tens of micrometers (in registration of γ-quanta and secondary x-rays)

of any of the previously mentioned types of radiation (Fig. 6.1).

The application of Mössbauer spectroscopy for the investigation of defects and radiation damage phenomena has been the subject of discussion in a number of studies, for example [351–355]. A detailed analysis of a large number of works concerned with the study of the implantation of iron ions by the nuclear gamma resonance method was published in [356]. The initial investigations of implanted impurities by means of Mössbauer spectroscopy were carried out in 1965 [357]. Implantation was carried out by means of recoil of Coulomb-excited nuclei of ^{57}Fe atoms (Coulomb excitation of the nuclei takes place as a result of the interaction of their self-electric field with the field of rapidly flying bombarding particles). The emission spectra, corresponding to the resonant transition of 14.4 keV, were measured in the transmitted γ-beam.

The doses investigated by this method are extremely small and equal $10^{10}\div10^{12}$ atom/cm^2, or further cm^{-2}. The implanted recoil ions are characterised by high energy (in [357] this energy was 30 MeV). The state of the atoms analysed in a very short period of time after implantation ($\approx$100 ns). This method of implantation is rarely used, because it requires the long operating time of the accelerator.

The implantation of the ^{57}Co mother isotope, which makes it possible to examine doses of $10^{12}\div10^{14}$ cm^{-2}, *i.e.*, the concentrations $10^{-4}\div10^{-2}$ at.%, was carried out for the first time in [358] using an isotope separator. The method is highly efficient in the investigation of the interaction of implanted impurities and defects. Its shortcomings are: strong radioactive contamination of the isotope separator (in the case of ^{57}Co, for example, $\tau_{1/2}$ = 270 days) and low implantation efficiency (1–5%).

The initial investigations of the stable isotope ^{57}Fe with the application of the isotope separator were carried out in 1973 by Savitsky *et al* (see the review [356] and the references in therein). In the initial investigations, measurements were taken of the transmission spectra for the implantation doses of $10^{16}\div10^{17}$ cm^{-2}. The Mössbauer spectroscopy of conversion electrons makes it possible to examine doses of the stable isotope starting from $5\cdot10^{13}$ cm^{-2}.

In the present chapter, analysis is made of the investigations of ion-doped metals and alloys mainly for the last 15–20 years, including the studies carried out by the author of the book and his colleagues using Mössbauer spectroscopy. Special attention is given to several problems which have not as yet been studied sufficiently in the literature. This relates to the theory-predicted possibilities and the results of experimental determination of the parameters of the rigidity of the interatomic bond (mean square of displacement of the atoms, Debye temperature), investigation of the different types of phase transformations, induced by ion bombardment and not associated with the processes of thermal and radiation-stimulated diffusion, in particular, amorphisation and martensitic transformations, and also 'long-range' effects: ion-induced structural phase transformations in subsurface layers of the materials with a depth of up to 10^4 R_p, taking place at relatively low (less than $10^{15}\div 10^{16}$ cm^{-2}) radiation doses in the absence of any significant heating of targets. In contrast to the destructive effects in the zone of the range of the ions, these transformations cause the transition of anomalously deep layers of metastable media to the state with a lower free energy.

6.2. EFFECT OF RADIATION DEFECTS ON THE PARAMETERS OF THE RIGIDITY OF INTERATOMIC BONDS

In [359], the authors investigated theoretically the problem of the effect of radiation defects, formed in irradiation of solids with fast neutrons or heavy ions, on the probability f (or f') of the Mössbauer effect (as already mentioned, the variation of the area below the resonance curve is proportional to the variation of product ff'). It was assumed that the zone of sensitivity of the given Mössbauer procedure (over the depth of the specimen) is characterised by the relatively uniform distribution of defect cascade zones. Calculations were carried out in continuous approximation. The strongly distorted regions of the type of 'displacement spikes' [335], formed as a result of the evolution of cascades of atomic collisions and characterised by the high density of radiation defects, were taken into account as microvolumes with a lower mass density.

It was assumed that the variation of the probability of the Mössbauer effect $f = \exp(-2W\ (\Theta_D))$ is determined by the dependence of the effective Debye temperature Θ_D on the variation of local density of the substance (Grüneisen approximation was used: $\Theta_D \sim \rho^{\Gamma}$; ρ is the density of the substance; Γ is the Grüneisen constant). It was concluded that the radiation defects should reduce the probability of the Mössbauer effect. The predicted relative reduction of the probability of the effect depends on the nature of radiation defects, the volume fraction of defect areas, and on temperature. The calculated decrease of f for highly defect α-Fe with Θ_D = 420 K is close to 1.5% at room temperature which is in agreement, in the order of magnitude, with the experimental data [360].

In addition to the formation of defects, ion implantation is accompanied by the enrichment of the surface layer by the implanted impurities (whose concentration in the subsurface layers may reach several tens of percent), redistribution of the inherent components of the target takes place in the vicinity of the irradiated surface (see paragraph 6.4), including the probable formation of different phases. That is why additional changes in the probability of the Mössbauer effect should be expected. For individual phases, formed during ion implantation, the difference in the probabilities f and f' may reach several percent [361].

6.3. LOCALISATION AND MOBILITY OF IMPLANTED ATOMS. THE STRUCTURE AND STABILITY OF FORMED RADIATION DEFECTS

Initial Mössbauer studies of implanted layers already showed the high information content of the method in respect of the determination of the areas of localisation and the parameters of electrical and magnetic interaction of the implanted atoms. In [362], the authors detected the existence of two states of the ions of ^{57}Fe, implanted in aluminium (Fig. 6.2). It was shown that the application of different Mössbauer techniques makes it possible to analyse the state of implanted Mössbauer nuclei selectively in depth [356] (Fig. 6.3). The presented transmission spectrum and the spectrum of conversion electrons greatly differ (in the first, there is a line associated with the nuclei of ^{57}Fe, situated at a larger depth).

In this paragraph, we present the data on the determination of the areas of localisation of implanted atoms, investigation of their mobility, identification of the type and stability of the resultant defects, obtained by the Mössbauer effect. In this case, we discuss mainly the implantation into the single-component targets (or, in some cases, strictly stoichiometric compounds).

In [363], transmission Mössbauer spectroscopy was used to determine the position by the implanted ^{57}Fe atoms in Al, Zr and Sn, and also the thermodynamic stability of different implantation positions. After Coulomb excitation, ^{57}Fe recoil atoms with the energy of 50 MeV were implanted in metal foils at temperatures of 5÷650 K. After implantation at low temperatures, the Mössbauer

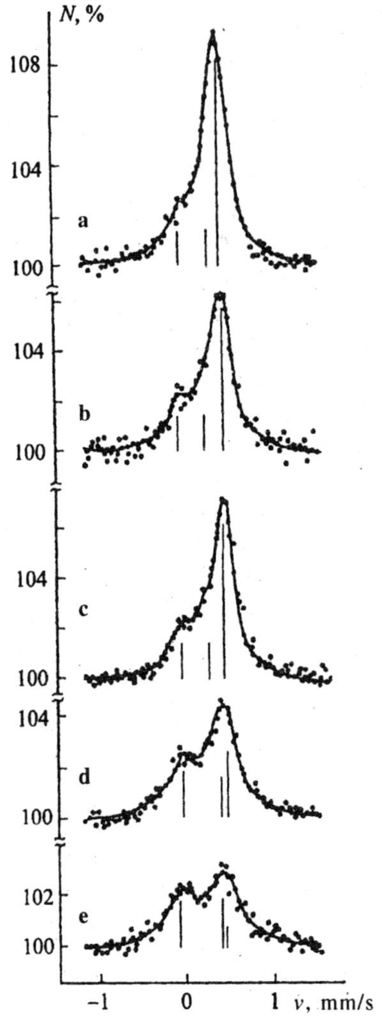

Fig. 6.2 Conversion electron Mössbauer spectra [362] of aluminium implanted with ^{57}Fe ions (fluence $2 \cdot 10^{15}$ cm^{-1}, with energies of *a*) 70 keV, *b*) 55 keV, *c*) 40 keV, *d*) 25 keV, *e*) 10 keV. The corresponding mean values of the concentration of ions in the surface layer with a thickness of $4\Delta R_p$ (ΔR_p is range straggling) are equal to c = 0.5; 0.6; 0.9; 1.1 and 2.1%. The lines indicate the positions of single peaks, corresponding to 'monomers' of iron and doublets, corresponding to 'dimers' of iron.

spectra of γ-radiation of the Coulomb-excited recoil atoms, showed two lines, one of which corresponded to the substitutional solid solution, and the other to the interstitial solution. With an increase of temperature the intensity of the second line decreased. This is explained by the high mobility of the implanted iron atoms in the interstices of metal matrices. The line corresponding to the interstitial solution was not detected above 250÷300 K.

The effect of disappearance of the subspectrum (quadrupole doublet), corresponding to the iron atoms, with the implanted carbon atoms located in the nearest interstices, was observed for the first time in 1968 by the authors of [364] for the iron–carbon austenite.

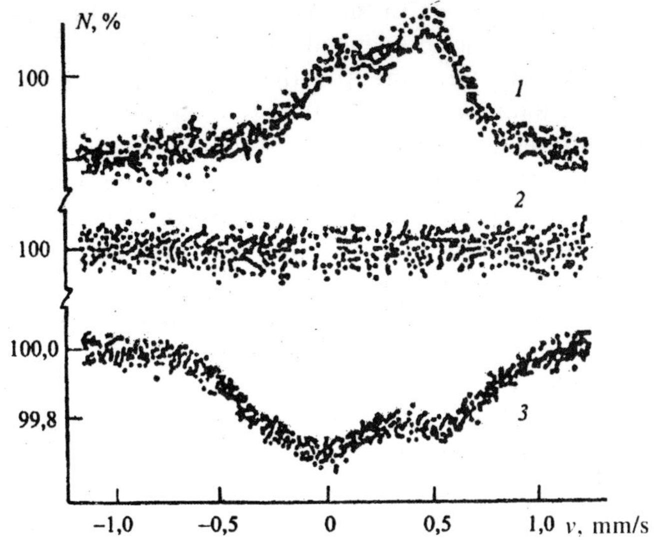

Fig. 6.3 The Mössbauer spectrum of conversion electrons (*1*) and transmission Mössbauer spectrum (*3*) of one and the same aluminium specimen implanted with ions of ^{57}Fe (fluence 10^{15} cm^{-2}, ion energy 35 keV, mean concentration ^{57}Fe 0.5%, thickness of Al foil 0.3 mm), and the Mössbauer spectrum of a non-implanted foil (*2*) [356].

Because the frequency of diffusion jumps of the interstitial carbon atoms sharply increases with temperature, the local atomic environment, averaged out over the lifetime of the excited state of the resonant isotope ^{57}Fe (~98 ns), becomes cubically symmetric for all Mössbauer nuclei, the quadrupole splitting in the Mössbauer spectrum of carbon austenite disappears. The distinction of the above situation is only in the mobility exhibited by the interstitial ^{57}Fe atoms themselves.

The authors of [365] investigated the effect of temperature on the parameters of Mössbauer resonance for ^{57}Fe atoms, implanted (after Coulomb excitation) into the interstices of the silicon single crystal. Measurements were taken in the temperature range 300÷850 K. The implantation of the fast ions of iron in silicon took place after Coulomb excitation of nuclei of ^{57}Fe by argon ions with the energy of 89 MeV. Temperatures higher than 550 K were accompanied by a large increase of the width of the resonance line which is explained by the increase of the rate of diffusion of the implanted atoms (see [347]). At temperatures of 700÷800 K, the excessive width of the line is close to 2 mm/s. The determined values of the diffusion coefficient are in agreement with the data obtained by other methods. The temperature dependence of the diffusion rate corresponds to the Arrhenius law, indicating that the mechanism of diffusion is independent of temperature. The isomeric shift of the Mössbauer line for the iron atoms in the silicon interstices at a temperature of 300 K is 0.84 mm/s.

In [366], Mössbauer emission spectroscopy was used to examine the localisation of radioactive atoms ^{57}Co after implantation in an aluminium single crystal. At room temperature, measurements were taken of the parameters of the hyperfine structure of Mössbauer spectra for 14.4 keV emission of daughter ^{57}Fe atoms. The spectra are characterised by the presence of a singlet line, corresponding to the formation of a substitutional solid solution, and also of a quadrupole doublet. The results show that the quadrupole splitting of the second component of the spectrum changes with the variation of the direction of the beam of the implanted atoms in relation to the crystallographic axes. The splitting was 0.21 mm/s in implantation along the [111] axis, and 0.15 mm/s in implantation along the [100] axis; this is explained by the localisation of impurity atoms in the defect sites of different type.

Annealed specimens of high-purity copper (99.999% Cu) were bombarded [361] with iron ions with the energy of 50 keV to doses of (0.3–6.5) · 10^{16} cm^{-2} at room temperature in the pulsed regime (10 μs, 5 mA/cm^2, 50 Hz). The authors paid attention to the fact that the ion current density was 3–4 orders of magnitude higher than the current used in the continuous implantation regime. After implantation, the material was investigated by the method of conversion electron Mössbauer spectroscopy. The effective thickness of the layer under study was 100 nm. The resultant spectra did not show precipitates of α-Fe in the implanted layers, whereas the mean concentration of Fe in Cu at doses of 10^{16} and $6 \cdot 10^{16}$ cm^{-2} was 4 and 11 at.%, respectively. The structure of Mössbauer spectra is linked with the existence of Fe atoms in the following states: in the form of single atoms, clusters, precipitates of γ-Fe, and also in the form of $FeCuO_2$ compound. It has been reported that in the regime of implantation of iron and copper, used in the given study, with the higher ion current density the number of the isolated atoms of iron is almost twice the number in the case of normal implantation, because of the stronger mixing of the components in the doped layer.

The effect of heat treatment on the supersaturated solid solution of Fe in Cu, obtained as a result of implantation of the Fe ions in Cu at a total dose of implanted atoms of 10^{16} cm^{-2}, was investigated by Mössbauer spectroscopy by the authors of [367]. Annealing at temperatures below 440°C results in gradual breakdown of the solid solution with the precipitation of α-Fe. This process is preceded by the breakdown of impurity-vacancy complexes which starts at temperatures of approximately 300°C. At temperatures higher than 520°C the solubility of Fe in Cu increases and this is reflected in partial dissolution of the α-Fe phase.

The state of the Fe atoms, implanted in Gd, was investigated by conversion electron Mössbauer spectroscopy [368]. Specimens were produced by implantation of $^{57}Co^+$ ions (50 keV, $2 \cdot 10^{13}$ cm^{-2}) and $^{57}Fe^+$ (70 keV, $5 \cdot 10^{15}$, 10^{16}, $2 \cdot 10^{16}$ and $4 \cdot 10^{16}$ cm^{-2}) in the polycrystalline Gd foils at room temperature. The doses of the implanted ions corresponded to the mean concentration of 0.01÷12 at.%. Implantation was accompanied by the local amorphisation of Gd. Immediately after implantation, the lines indentified in spectrum corresponded to the isolated atoms of Fe, dimers Fe_2 and atoms Fe bonded with the oxygen atoms. In the course of annealing at 400÷500 K the processes of structural relaxation took place. It is assumed that at 500 K and higher there occur the processes associated with interstitial ordering of the iron atoms and/or formation of intermetallics. Annealing at a temperature of 700 K in the presence of oxygen is accompanied by the formation of an amorphous oxide film.

It is of interest that in the case when the atoms of the metal–target and of the implanted impurity have a strong tendency towards the formation of an ordered solid solution (because of high ordering energy), the process of atomic ordering in the course of ion bombardment may dominate over destructive processes, leading to the formation of defects and chaotisation. Thus, Reuther [369] used conversion electron Mössbauer spectroscopy to study pure iron, implanted with different doses of aluminium ions ($5 \cdot 10^{16} \div 5 \cdot 10^{17}$ cm^{-2}) with the energy of 50, 100 and 200 keV. Regardless of the fact that the doses exceeding $1 \cdot 10^{17}$ are sufficient for producing an absolutely disordered layer, depending on the dose and energy of the implanted ions, there was observed the short-range atomic order of the type Fe_3Al or FeAl. In annealing in the temperature range 400÷600 °C, the short-range order of the type FeAl changes to the short-range order of the type Fe_3Al. At annealing temperatures of approximately 700°C, the main phase recorded by conversion electron Mössbauer spectroscopy is the non-ordered solid solution of aluminium (several at.%) in iron.

Mössbauer isotope ^{57}Co [370] was introduced to a mean concentration of 0.001 at.% into molybdenum specimens of high purity (99.999%) by the diffusion method. The diluted alloy MoCo, produced by this method, was irradiated with protons with the energy of 70 keV to a flux of $2 \cdot 10^{18}$ cm^{-2} at a temperature of 300 K and subsequently subjected to isochronal annealing in the temperature range 300÷1300 K, with measurements taken of the γ-resonance spectra during annealing. In addition to the single line, corresponding to the Co atoms in the substitutional position, investigation showed six additional components, five of which (2–6) were represented by symmetric doublets with the initial width, and component 1 by a single widened line. The component 1 is attributed to the complexes: (intrinsic interstitial atom)–(Co atom). Components 2–4, formed as a result of annealing in the temperature range 400–500 K are associated by the authors with the complexes (vacancy)–(Co atom), and the components 5 and 6, with the nonequilibrium segregations of Co atoms.

Investigations were carried out into the effect of proton irradiation on the parameters of the Mössbauer resonance line 14.4 keV of γ-radiation of impurity atoms of ^{57}Fe in platinum [371]. The specimens were irradiated with protons with the energy of 1 MeV at 90 K and annealed at temperatures in the range 100–1420 K. The spectrum of the non-irradiated specimens presented a single line with the natural width. After irradiation, the spectra contained additional ('defect') lines whose relative intensity depends on the radiation dose and annealing temperature. Defect lines represent quadrupole doublets with a splitting of 0.18–0.26 mm/s. The authors discuss the features of interaction of the impurity atoms with radiation defects in platinum at different temperatures.

The possibilities of conversion electron Mössbauer spectroscopy may also be illustrated by the data presented in [372], in which the authors investigated the effect of implantation of hydrogen H_2^+ (energy 20÷80 keV, dose $1 \cdot 10^{16} \div 4 \cdot 10^{16}$ cm^{-2}) in a single crystal of yttrium ferrogarnet on the parameters of hyperfine interaction of Fe^{3+} ions. The implantation of hydrogen was followed by the broadening of resonance lines and by a decrease of the values of the local magnetic fields, acting on the nuclei of Fe^{3+} ions. Particularly, a large variation of the parameters of a hyperfine interaction was detected for the ions in tetrahedral positions. Approximately 15% of implanted hydrogen ions form stable chemical bonds with oxygen ions leading to the formation of a new non-equivalent position of the Fe^{3+} ion (and of the corresponding Mössbauer subspectrum). The implantation of hydrogen is also accompanied by a change in the orientation of magnetic moments of the iron ions. Measurements were taken of the dependence of the parameters of the hyperfine structure on the annealing temperature of implanted specimens. The results were compared with the similar data, obtained previously for yttrium ferrogarnet, implanted with neon ions.

In [373], conversion electron Mössbauer spectroscopy was used to investigate the effect of the ion current density (j = 0.1÷1.5 μA/cm^2) in the implantation of Ne^+ ions (E = 100 keV) in the epitaxial ferrite–garnet films on the concentration of the resultant radiation defects and the distribution of the defects over the depth of the irradiated layer. The experimental results show that the increase of the current density results in the formation of complicated complexes of defects which play a significant role in the formation of the properties of irradiated films.

Emission Mössbauer spectroscopy [374] was used to examine the state of ^{119}Sn atoms formed as a result of the radioactive decay of the atoms of ^{119}Sb, implanted using an isotope separator into a number of metal matrices. The spectra contained components corresponding to substitutional solid solutions, defect nodes and, possibly, intermetallic phases. In the case of formation of substitutional solid solutions, the isomeric shift in the matrices of Al, Y, β-Sn, Pt, Au, and Pb was found to be equal to respectively 2.36; 1.86; 2.51; 1.40; 2.52 and 3.25 mm/s. The results are discussed together with similar data for the impurity atoms of Sn in other metal matrices. Special attention was given to discussing the relationship of the value of the isomeric shift and parameters, characterising the matrix, and also the interaction of impurity atoms and the atoms of the matrix.

The relationships governing the changes in the defect structure of the elemental and phase composition of the surface layers of α-Fe after implantation of the ions of Ti, Al, C ($5\cdot10^{16}$÷$8\cdot10^{17}$ cm^{-2}) were investigated using the beam of slow positrons and conversion-electron Mössbauer spectroscopy [375]. The maximum depth of the layer with increased concentration of the defects did not exceed 300–350 nm and was comparable with the maximum range of the ions. The formation of the phases TiC, Fe_3Al, cementite and ε-carbide was detected. The largest effect of the increase of microhardness, the decrease of wear and friction coefficient was recorded for titanium, and a slightly smaller effect, for aluminium. The implantation of carbon only increases microhardness, without affecting wear resistance and friction coefficient.

Conversion electron Mössbauer spectroscopy was also used in [376] to examine the surface of polycrystalline foils of α-Fe, irradiated with ions of Ar^+ (E = 5 keV, F = $1 \cdot 10^{17}$ cm^{-2}), T = 36÷300 K. The distortion of the form of the Mössbauer spectra (the formation of the distribution of hyperfine fields) is linked with the damage of the surface region by the ions of Ar^+. Immediately after implantation, the profile of distribution over the depth of the components of the spectrum, associated with damage, showed a maximum on the irradiated surface (>55%) and extended to a depth of >20 nm. This greatly exceeded the calculated depth of penetration of the Ar^+ atoms.

The non-equivalent positions of ^{57}Co, implanted in Si and Co_xGa_{1-x} can be detected by the Mössbauer effect [377, 378]. It has been shown that ^{57}Co, implanted in Si, is trapped by voids into positions Q_1 and Q_2, differing in the binding energy. At high temperatures, silicide formation is detected in the initial stage, and with time at temperatures above 750°C the particles of the silicide partially dissolve and capture in positions Q_1 becomes the dominant process. In [378], the broadening of the spectra of the specimens of Co_xGa_{1-x} (x = 0.45÷0.58), implanted with the ions of ^{57}Co in comparison with the specimens into which ^{57}Co was introduced by diffusion, is explained by the decrease of the local symmetry of ^{57}Co because of the formation of vacancies and the increase of the number of Co atoms in antistructural positions of the lattice of the CsCl type. After annealing at T = 800°C the spectra were the same as for the non-implanted specimens.

6.4. VARIATION OF THE COMPOSITION OF THE SURFACE LAYER OF MULTI-COMPONENT TARGETS

Transmission Mössbauer spectroscopy was used for the investigation [340, 379] of foils (30 μm) of $Fe_{65}Al_{35}$, initially ordered in accordance with the type $B2$ by annealing at 650°C (1 h), and irradiated after that with the N^+ ions (E = 20 keV, j = 50 μA/cm^2). At the radiation doses stronger than $5 \cdot 10^{17}$ cm^{-2}, the structure of the spectrum showed lines of pure α-Fe (Fig. 6.4). The data obtained in secondary ion mass spectrometry indicate separation of the surface layer of the alloy into sublayers with a thickness of approximately 50÷100 nm consisting of almost completely pure Fe and Al and containing nitrogen implanted in the interstices and, according to the data of ESCA, a certain amount of the oxides of Fe and Al. This effect differs qualitatively from the available data

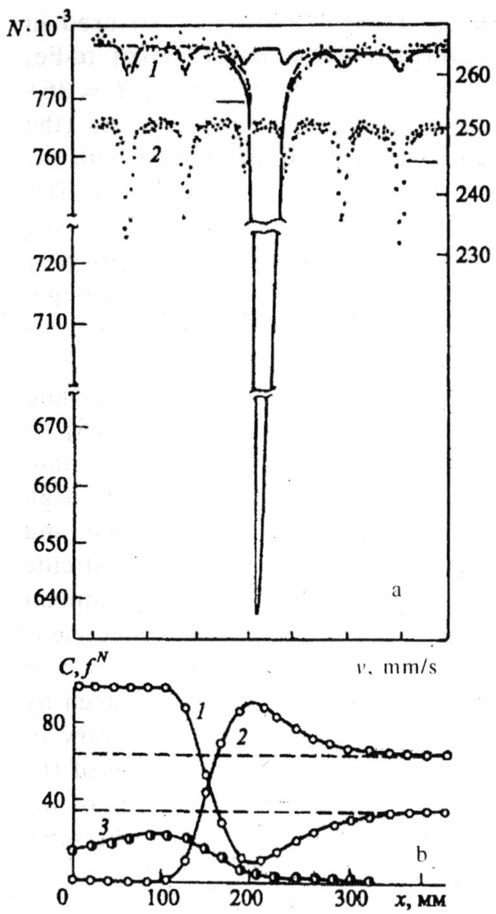

Fig. 6.4 Results of investigation of the $Fe_{65}Al_{35}$ alloy [340]: *a*) Mössbauer transmission spectra of alloy prior to (dashed line) and after (solid line) ion bombardment (*1*), and also the spectrum of pure ion (*2*); *b*) distribution of components in the surface layer: $C(x)$ is the atomic concentration of the atoms of Al (*1*) and Fe (*2*); $f^N(x)$ is the ratio of the number of atoms N to the total number of nodal atoms of Fe and Al (*3*).

on the partial separation of Fe–Al and other alloys in irradiation with neutrons and also ions and electrons [380] and is evidently caused by the difference in the partial coefficients of diffusion of Fe and Al by both the vacancy and interstitial mechanisms[1] which in movement of the defects, reproduced by irradiation, to the sinks results in the formation of diffusion counterflows of Fe and Al of different strength (the phenomenon of the type of reverse Kirkendall effect [380]). Consequently, long-term radiation may result in a very large difference of the concentration of the components in the immediate vicinity of the sink and at some distance from it, up to the separation of the alloy into pure

[1]For example, if a vacancy tends to migrate towards the surface over the 'iron' nodes, then iron will be pulled into the depth of the target, and aluminium will enrich the surface.

components. The increase of the diffusive mobility of the atoms is aided by in heating of the specimen under long-term high-current ion irradiation.

6.5. FORMATION AND DISSOLUTION OF PHASES, AMORPHISATION

In [381], conversion electron Mössbauer spectroscopy was used for the investigation of the effect of the radiation dose with the B^+ ions on the structure of polycrystalline films of α-Fe with a thickness of 1500 Å. When the dose was slightly higher than 10^{17} cm^{-2}, an amorphous phase formed in the film. Its amount increased with an increase of the radiation dose. The hyperfine field and isomeric shift remained constant until the entire film became an amorphous alloy $Fe_{75}B_{25}$ (at a dose $\sim 4\cdot 10^{17}$cm^{-2}). Annealing for one hour at 150°C resulted in the precipitation of α-Fe. After holding for 1 h at 300°C, the amount of the amorphous phase greatly decreased, indicating diffusion of B into α-Fe. Annealing for one hour at 400°C resulted in complete crystallisation of the film with the formation of a mixture of phases α-Fe and Fe_2B.

The methods of high-voltage transmission electron microscopy and conversion electron nuclear gamma resonance were used to investigate the implantation of the ions of B^+, N^+ and P^+ in iron and different steels [382]. Polycrystalline specimens were irradiated at room temperature with ions with the energy of 50–100 keV. The radiation dose was varied in the range $10^{16} \div 10^{18}$ cm^{-2}. Depending on the type of implanted ions and irradiation conditions, the surface layer showed formation of different compounds and/or amorphisation. The irradiated specimens were annealed at temperatures up to 600°C. Measurements of the microhardness and abrasive properties of the specimens, irradiated with nitrogen ions, showed that the optimum energy of the ions is in the range 40÷80 keV. In this case, the thickness of the implanted layer is 100÷200 nm. The optimum radiation dose is $(4 \div 8)\cdot 10^{17}$ cm^{-2} and the optimum density of ion current is 5÷8 μA/cm^2. At higher service temperatures all positive effects from implantation disappear because of the instability of the ion-induced effects and precipitates of the secondary phases.

The specimens of Fe [383] were bombarded with ions of B^+, C^+ and P^+ with an energy of 50 keV to a dose of 10^{18} cm^{-2}. Conversion electron Mössbauer spectroscopy was employed to investigate the amorphous layers formed on the surface and the variation of the state of the layers after vacuum annealing at 300÷700°C. The implantation of B^+ results in the formation of a layer of a disordered Fe–B alloy which transforms to the FeB compound after annealing at 400°C and to Fe_2B at 500°C. The implanted layers of the Fe–C alloy corresponded, in respect of the local atomic structure and composition, to the compounds Fe_5C_2 and Fe_2C. After annealing at 300÷400°C, the amorphous alloy crystallised with the formation of the Fe_5C_2 compound which, after annealing at 500°C, transformed to Fe_3C. The implantation of P^+ ions resulted in the formation of the Fe_2P amorphous layer with paramagnetic properties. Annealing at 500°C resulted in the formation of Fe_3P compound.

In [384, 385], the methods of conversion Mössbauer and Auger electron spectroscopy were used to examine the elemental and phase composition and also the local atomic structure of thin iron films, subjected to successive implantation of boron and nitrogen ions. It has been established that the implantation of the boron ions results in the formation of magnetic and paramagnetic amorphous phases α-(Fe–B). Subsequent irradiation with nitrogen was accompanied by the breakdown of the α-(Fe–B) magnetic amorphous phase, with the formation of compounds of the type of boron nitride. The short-range order in the resultant amorphous structures in relation to the concentration of B and N was studied. The results show that the short-range order in the Fe–B–N system is determined by the concentration of implanted boron.

Conversion electron Mössbauer spectroscopy [386] was used to investigate thin layers of α-Fe after implantation of the N_2^+ ions (100 keV, $5 \cdot 10^{17}cm^{-2}$), N^+ (50 keV, $10^{18}cm^{-2}$) and P^+ (50, 100 and 200 keV, $1 \cdot 10^{17}$ cm^{-2} and $5 \cdot 10^{17}$ cm^{-2}). The resultant conversion and Auger electrons with the maximum energy of 7.3 keV and the yield depth of 300 nm were recorded. The main part of the Mössbauer signal (~65%) is related to the layer with a thickness of 50 nm which is most interesting for examining ion implantation. In the specific conditions of nitrogen implantation, investigation showed formation of iron nitrides. The implantation of the P^+ ions was accompanied by both the amorphisation of the implanted layer and formation, in the layer, of iron phosphides of different composition. The transformations of implantation-induced phases in the process of isochronous annealing in the temperature range 400÷700°C were investigated.

The methods of conversion electron Mössbauer spectroscopy, x-ray diffraction and Auger electron spectroscopy [387] were used in investigation of subsurface layers of thin (40÷100 μm) foils of iron, enriched with the isotope ^{57}Fe to 95%, after implantation of the nitrogen ions with the energy of 45 keV and of carbon ions with the energy of 40 keV up to doses of $5 \cdot 10^{17}$ cm^{-2} and subsequent isochronous annealing lasting for 30 min in the temperature range 200÷700°C. After carbon implantation, the Mössbauer spectrum contained (in addition to the matrix) the Fe_2C metastable phase. It has been shown that θ-Fe_3C cementite forms only as a result of annealing at temperatures higher than 200 °C. After implantation of nitrogen, the analysed layer contained the ε-Fe_4N nitride. Annealing in the temperature range 300÷400°C resulted in the formation of the γ-Fe_4N phase. The results of investigation of the previously detected phenomena of 'natural ageing' of specimens irradiated with nitrogen ions are presented.

Thin films of ^{57}Fe, deposited on SiO_2 [388], were implanted with ions of $^{64}Ni^+$ isotope with the energy of 80 keV and doses of 0.5÷ $10 \cdot 10^{16}$ cm^{-2}. The structure was investigated by the methods of Rutherford backscattering and conversion electron Mössbauer spectroscopy. The ion-implanted layers contained in different proportions (in relation to the implantation dose) the Fe–Ni BCC phase, enriched with iron, the FCC phase, enriched with nickel, and Fe–Ni invar alloys with the FCC structure. With an increase of the implantation dose the content of the BCC Fe–Ni phase decreases and the content of the phases characterised by the invar behaviour increases.

In [389], the method of Mössbauer spectroscopy was used to investigate the zones of radiation damage in the vicinity of primary knocked out atoms of ^{57}Co (^{57}Fe) in specimens of Ni_2Fe alloy, irradiated with protons with an energy of 30 MeV and α-particles with energies of 40 and 50 MeV. Analysis of the spectra shows that the damaged regions may be characterised by the formation of precipitates of a new phase (for example, NiFe) enriched with iron. These precipitates grow during holding at room temperature. This is accompanied by the ordering of the alloy as a result of radiation-enhanced diffusion. The type and energy of the irradiated particles have no effect on the investigated properties of the alloy, and the only controlling parameter is the level of damage in the specimen.

In [390], conversion electron Mössbauer spectroscopy was used to explain the nature of the effect of bombarding particles in the secondary radiation of implanted steels on the resultant carbonitrides. In continuation of previous experiments with α-particles (He^+), steel 1020 with precipitates of carbonitrides, formed as a result of implantation of N^+ ions, was irradiated with Ar^+ ions with the energy of 150 keV, dose of $1 \cdot 10^{14}cm^{-2} \div 1.5 \cdot 10^{16}cm^{-2}$, with the temperature of the specimen controlled at 25°C. Investigation shows the dual effect of irradiation with argon ions: 1) precipitates of both ε-Fe_2(C, N) and ε-Fe_{2+x} (N, C) dissolve completely or partially (for comparison, the α-particles resulted only in the dissolution of the carbonitrides of the second type); 2) the presence of the Ar atoms changes the thermal characteristics of the carbonitrides, increasing the dissolution temperature of the carbonitrides in subsequent annealing from 400 to 450°C (in this respect, the effect of Ar is also more efficient than that of α-particles).

Investigations were carried out on 1020 steel [391], containing 0.2% C and 0.9% Mn, after initial implantation of nitrogen ions and subsequent implantation of xenon ions. The energy of the implanted N_2^+ ions was 40, 80 and 140 keV, the depth of penetration of the ions 2000 Å, irradiation dose $6 \cdot 10^{16}\ N_2^+ / cm^2$, the concentration of nitrogen atoms in the surface layer reached 40 at%. The energy of the Xe^+ ions was 500 keV, the depth of penetration 650 Å, the radiation dose varied in the range $10^{14} \div 10^{16}$ cm^{-2} To avoid heating of the specimens, the density of current in implantation of nitrogen and xenon was 1 μA/cm^2. In addition to this, the specimens were cooled, maintaining the temperature equal to 20°C. The results were analysed by the CEMS and nuclear reactions methods. The results show the dissolution and partial precipitation of carbonitrides. The solubility temperature of the carbonitrides in the presence of the Xe^+ ions in the implanted layer increased. The results were compared with the data obtained previously in bombardment of 1020 steel with the ions of He^+ and Ar^+. Bombardment with the Ar^+ ions resulted in the highest stability of the carbonitrides in subsequent annealing at 450 °C. Bombardment of the surface with Xe^+ ions, resulting in the same displacement per atom as He^+ and Ar^+, has a stronger effect on the processes of dissolution and secondary precipitation of the carbonitrides. The effect of the type of bombarding ions on the stoichiometric composition of carbonitrides has been established.

Conversion electron Mössbauer spectroscopy was used to examine the structural-and-phase state of the surface of α-iron, irradiated with oxygen ions [392]. The following sequence of phase transformations in the ion-doped layer was found:

$$Fe \xrightarrow{O^{+}(50\,keV)} Fe_2O_3 + Fe_3O_4 \xrightarrow{300\,°C} Fe_3O_4 \xrightarrow{600\,°C}$$
$$\rightarrow Fe_3O_4 + FeO \xrightarrow{700\,°C} Fe_3O_4 + FeO + Fe.$$

The same authors investigated [393] the state of the surface of α-Fe irradiated with boron ions. Conversion electron Mössbauer spectroscopy was used. The results show a sequence of phase transformations after implantation and annealing.

The authors of [394–396] investigated different aspects of the formation of the phases ε-FeSi, α- and β-$FeSi_2$ in the implantation of Fe (enriched with ^{57}Fe) in the *n*-Si by the methods of Mössbauer and Auger electron spectroscopy. The results show that, depending on the implantation conditions, these phases or their mixtures form. It has been shown in [396] that in implantation of Fe^{+} with an energy of 200 keV at 350°C (fluence $7 \cdot 10^{17}$ cm^{-2}) only the β-$FeSi_2$ phase forms in the vicinity of the surface, and a mixture of ε-FeSi and β-$FeSi_2$ forms at a large depth. It has been shown that these phases form during implantation at concentrations lower than those indicated by the equilibrium phase diagram. The Mössbauer spectroscopy results [395] indicate that either the α- or β-phase may form from the mixtures of α- and β-phases, depending on annealing temperature.

Mössbauer spectroscopy was also used to examine [397] silicides produced by implantation of ^{57}Fe (100 keV) in Si (100) at room temperature with subsequent irradiation with a beam of Si ions (500 keV) at 593 K to initiate epitaxial crystallisation. The spectra of the implanted specimens are characterised by a wide distribution of quadrupole splitting. After annealing at 793 K this distribution became narrower and the spectrum parameters approached those of β-$FeSi_2$.

In [398], the methods of conversion electron Mössbauer spectroscopy and scanning Auger electron spectroscopy were used to investigate the structure of Si (111) of the *n*-type after implantation of Fe^{+} ions (200 keV, $3 \cdot 10^{17}$ cm^{-2}, 350 °C) and subsequent annealing (900 °C, 18 h, 1150 °C, 1 h). The experimental results show that the mixture of the β- and α-$FeSi_2$ phases, formed after implantation, completely transforms to the β-phase during annealing at 900 °C, whereas annealing at 1150 °C results in the complete β→α transformation. The method of scanning Auger spectroscopy showed concentration inhomogeneities, not known for the Fe–Si system (absent according to the equilibrium phase diagram). In the iron-depleted matrix (less than 2 at.% Fe) there is the formation of regions enriched with iron to approximately 30 at.% where there appear either large precipitates β-$FeSi_2$ (900 °C) or a network structure β-$FeSi_2$ (1150 °C).

In [399], detailed investigations were carried out into the possibility of amorphisation of nonequilibrium bulk (produced by ion mixing) and thin-film multilayer systems Fe/Al by the methods of x-ray diffraction, CEMS, EXAFS-spectroscopy and electron microscopy. The experimental results obtained for homogeneous bulk specimens, produced as a result of mixing of Fe/Al multilayers with the ions of $^{56}Fe^+$ (150 keV) show that at a content of aluminium of up to 70 at.%, a BCC solid solution forms in the material. The concentration region in the vicinity of 75 at.% Al is characterised by the formation of the Fe_2Al_5 crystalline phase. The amorphous state with the short range order Fe_2Al_5 was found at 80 at.% Al. In the thin films (Fe/Al multilayers), the amorphous state in the region of the interface may exist at a lower content of aluminium (~70 at.%) because of the high energy of the transition layers (interfaces). Ion mixing at 140 K may reduce the concentration in amorphisation to ~60 at.% Al because of kinetic blocking of the crystallisation process.

In [400, 401] the methods of CEMS, Auger electron spectroscopy and x-ray diffraction were used to examine Fe–Mg and Fe–Al alloys produced by implantation of the Fe^+ ions (195–200 keV, $1\cdot10^{14}$–$9\cdot10^{17}$ cm^{-2}) in magnesium and aluminium. The maximum concentration of Fe reached 60 at.% in the case of aluminium and 94 at.% for magnesium. At a fluence not higher than $1\cdot10^{17}$ cm^{-2}, the Mössbauer spectra for both systems have the form of paramagnetic line doublets, indicating the presence of at least two different states of the Fe atoms. At a fluence of $2\cdot10^{17}$ cm^{-2}, irradiated magnesium showed the formation of a ferromagnetic phase corresponding to the maximum concentration of the implanted iron atoms of approximately 60 at.%. However, in the aluminium target, the magnetic phase forms only at a flux of $9\cdot10^{17}$ cm^{-2} (c_{max} ~75 at.% Al). X-ray diffraction showed the reflections of FeAl, Fe_3Al and Fe in Al. The formation of amorphous phases was not detected.

In contrast to the alloys investigated above and produced by the implantation of Fe^+ in Mg and Al, the authors of [402, 403] investigated the alloys produced by implantation of Al^+ and Si (50 keV) in iron. It has been established that, depending on the fluence, as in the cases described previously, it is possible to produce magnetic and non-magnetic atomically disordered phases. The nonmagnetic disordered phase FeAl is unstable and ages with the formation of the ordered *B*2 phase already at 300 °C. Formation of the DO_3 phase was not reported.

The authors of [404] proposed a procedure for decoding conversion electron Mössbauer spectra which makes it possible to carry out quantitative phase analysis and reconstruct the profiles of distribution of the amount of phases over the depth of the specimens of α-Fe. The results of application of this procedure for analysis of high-dose implantation of nitrogen into iron are presented. The method shows that, regardless of the regions supersaturated with nitrogen, microvolumes of α-Fe remain in the surface layer.

6.6. ION-INDUCED MARTENSITIC TRANSFORMATION

In [405] the methods of transmission electron microscopy and conversion electron Mössbauer spectroscopy were used to investigate the mechanism of stimulation of martensitic transformations in austenitic stainless steels in the course of implantation. 17/7 austenitic stainless steel was implanted with ions of inert gases Kr^+ or Ar^+, or ions of intrinsic elements Fe^+, Cr^+, Ni^+. The ion energy selected was such that the depth of penetration was 40 nm. The form of the conversion electron Mössbauer spectra (Fig. 6.5) and also the data obtained by transmission electron microscopy indicate that irradiation of the steel with ions of Ni, Fe or Cr result in the formation of α'-martensite in the surface layer, regardless of the fact that the chemical effect of these elements differs: nickel stabilises the γ-phase, iron stimulates the formation of α'-martensite, and alloying with chromium should result in the formation of α-ferrite in addition to precipitates of the γ-

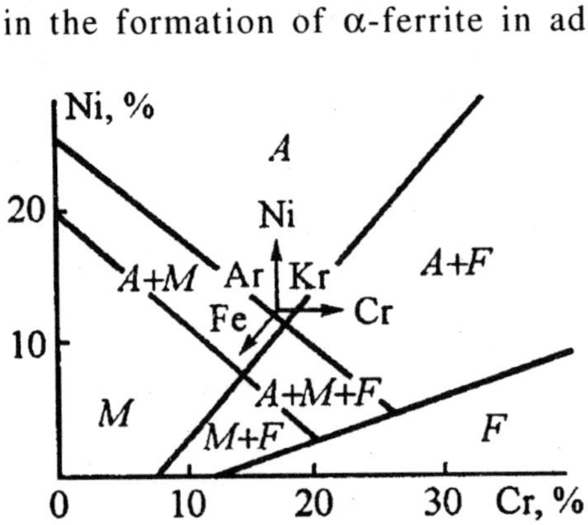

Fig. 6.5 Schäffler diagram (phases formed in Cr–Ni stainless steel in relation to the concentration of components) [408, 411]. Arrows indicate the variation of the composition of the surface of 17/7 steel in implantation of 'intrinsic' elements Fe, Cr, Ni to a fluence of $2 \cdot 10^{16}$ cm^{-2}, which corresponds to the mean implanted concentration in the layer with a thickness of $4\Delta R_p$ of 6 at.%.

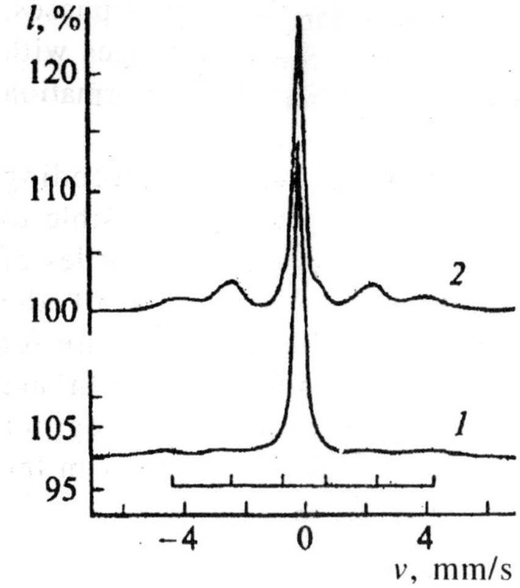

Fig. 6.6 Conversion electron Mössbauer spectra [408, 411] of 17/7 austenitic steel (18.4% Cr, 7.0% Ni and 74.6% Fe), implanted with 150 keV Fe to a fluence of $2 \cdot 10^{16}$ cm^{-2} (*1*) and 90 keV Ar to a fluence of $5 \cdot 10^{16}$ cm^{-2} (*2*).

phase (Fig. 6.6). It was concluded that the formation of martensite is associated mainly with stress fields from secondary radiation damage. This conclusion has been confirmed by the implantation of ions of inert gases at which the highest degree of martensitic transformation (approximately 100%) was obtained in the case of doses leading to the formation of bubbles.

In [406] the methods of conversion electron Mössbauer spectroscopy and x-ray diffraction were used to investigate phase transformations in Fe–18Cr–7Ni austenitic stainless steel, implanted with ions of Fe, Ni and Sn with energies of 150÷300 keV and doses of $10^{16} \div 10^{17}$ cm^{-2}. The experimental results show that different types of metallic ions provide different contributions to the ($\gamma \to \alpha'$)-transformation. It is shown that the ($\gamma \to \alpha'$)-transformation in Fe–18Cr–7Ni steel, implanted with ions of Fe and Ni, takes place under the effect of changes in the composition (doping effect) and stresses, induced by radiation damage (effect of stresses), and in the case of implantation of the steel with the Sn ions the driving force of the ($\gamma \to \alpha'$) transformation is determined only by the effect of stresses.

In [407] the methods of conversion electron Mössbauer spectroscopy and x-ray diffraction with grazing of the primary x-ray beam were used to examine changes in the structure of an austenitic stainless steel (18.4% Cr, 7.0% Ni) after implantation of ions of Xe^+, Kr^+, Ar^+, Fe^+ and Ni^+. The energy of the ions was 90÷300 keV, and the dose $2 \cdot 10^{16} \div 10 \cdot 10^{17}$ cm^{-2}. It was concluded that the relationships governing phase formation in implantation of the metallic ions are determined by doping and formation of stresses; the phase transformations in implantation of the ions of inert gases are determined only by the formation of stresses. An increase of the dose of implanted Fe^+ ions from $2 \cdot 10^{16}$ to $1 \cdot 10^{17}$ cm^{-2} reduces the amount of retained austenite from 22 to 13 conventional units and increases the amount of the α'-phase from 2 to 10 conventional units.

The methods of conversion electron Mössbauer spectroscopy and x-ray diffraction in grazing beams [408] were used to examine the martensitic transformation in Fe–19Cr–10Ni–0.08C steel, implanted with the ions of He, H and D with an energy of 8 keV. In addition to this, the distribution of martensite over the depth of the implanted layer was determined by energy-dispersion conversion electron Mössbauer spectroscopy. The martensitic transformation in the implanted layer takes place after implantation of 10^{17} cm^{-2} He ions, whereas the fluence required after implantation of H or D to reach the same effect is up to 100 times greater. The difference is caused by the capacity of He to form high-pressure gas bubbles, whereas implanted hydrogen is continuously lost as a result of reverse diffusion to the surface. The bubbles of He, subjected to a pressure of >60 GPa, induced extremely high stresses in the implanted layer which directly undergoes the martensitic transformation. The fact that considerably larger fluences of H or D are required to induce the martensitic transformation indicates that radiation damages proper play a secondary role in this case.

In [409] the methods of Mössbauer spectroscopy, x-ray diffraction and resonance nuclear reactions were used to investigate phase transformations in chromium–nickel austenitic stainless steel, implanted with nitrogen ions. Processes of formation of martensite in the austenitic matrix and subsequent transition back to austenite were observed. The results show that the limit of solubility of nitrogen in steel in ion implantation is exceeded and, at the same time, nitrides precipitate at relatively low nitrogen concentrations. The wear resistance of the surface may increase or decrease. The conditions of occurrence of $(\gamma \leftrightarrow \alpha')$-transformation in direct and reverse directions were discussed. The effects associated with the radiation nature of treatment and with the introduction of impurities have been investigated.

In [410], the method of conversion electron Mössbauer spectroscopy with successive etching of layers was used to investigate the structure in $Fe_{70}Ni_{27}Mn_3$ disordered alloys after proton radiation with an energy of 13 keV. The dose of implanted particles H^+ varied in the range $1.7 \cdot 10^{17} \div 2.3 \cdot 10^{19}$ cm^{-2}. The transition from the austenitic FCC phase to the martensitic BCC phase at a depth of 150÷250 nm was found. The authors also related the occurrence of phase transitions to high internal stresses, developed in the surface layer of the alloy during ion implantation which relax during transformation.

In [411] the authors carried out a detailed analysis of the phenomena of amorphisation and martensite-like transformations in metals and alloys during ion bombardment. In addition to ($\gamma \leftrightarrow \alpha'$)-transformations in stainless steel, investigation showed the occurrence of diffusion-free transformation FCC (α) $\rightarrow$ HCP (ε) in pure nickel in implanting small doses of phosphorus ions. Further implantation of phosphorus resulted in the amorphisation of the surface layer. The author related the amorphisation and martensite-like transformations to a single group of ion-induced transformations. The latter may be caused by ion bombardment in thermally diffusion-free conditions in which the effects of radiation-enhanced diffusion are also minimised. In this case, the controlling role in amorphisation is played by the chemical effects required for maintaining the amorphous state (as in conventional technologies of superfast quenching). The special role in the induction of martensitic transformations is played by static stresses, determined by the implantation of impurities. The effect of the latter is superposed with the chemical effect of the implanted elements. According to the results of [411], the martensitic transformation propagates during the ion-induced phase transformation to a depth 2–3 times greater than the projected range of the ions.

6.7. ION MIXING

The method of internal conversion electron Mössbauer spectroscopy [412] was used to investigate phases formed during irradiation of Fe/Mo multilayer films by ions of Xe^+ with an energy of 300 keV at a temperature of 77 K. The ferromagnetic phases, associated with the partial spectra with the hyperfine magnetic structure, are iron-based solid solutions with a variable molybdenum concentration. Two other phases in the spectra correspond to quadrupole doublets with the splitting of 0.09÷0.30 mm/s; the composition of these phases has not been determined. The possibility of formation of metastable amorphous alloys in the investigated system has been discussed. The results have been compared with similar data obtained for Fe–Si and Fe–Sn alloys.

Iron films and foils and also the binary films B/Fe (65 nm B on the substrate of α-Fe, enriched with ^{57}Fe) were irradiated with the ions of B^+ with the energy of 30 and 60 keV and ions of Ar^+ with the energy of 120 keV [413]. Investigations were carried out both by the method of the transmission Mössbauer spectroscopy and in the geometry of backscattering with registration of internal conversion electrons. The results show that irradiation is accompanied by the formation of amorphous phases whose composition is similar to the composition of alloys produced by deposition on cooled substrates. A decrease of target temperature to the liquid nitrogen temperature increases the amount of the amorphous fraction. In irradiation of the binary films with the ions of Ar^+, the mean boron concentration in the amorphous phases is higher than in the case of irradiation with B^+. The crystallisation of the amorphous phases starts at a temperature of 373 K and is compl-eted at 573 K. The main phase, formed in crystallisation, is Fe_2B.

The relationships governing the ion mixing of multilayer systems in relation to the composition and structure of targets, the type of ions and in irradiation temperature were also investigated in [414–417].

In [418] Mössbauer spectroscopy was used to investigate multilayer structures Fe/Tb at 300 K after bombardment with Xe ions. Investigation showed two different effects: conventional ion mixing and separation of the atoms of Fe and Tb at the interfaces. The first process rapidly disrupts the laminated structure and at a high dose it results in the formation of an amorphous alloy. The second process is detected only at small doses and, according to the authors, is associated with the thermal instability of these multilayer structures (relaxing to the state with minimum energy).

The methods of Rutherford backscattering and Mössbauer spectroscopy [419] were used for analysis of thin films of ion on a silicon substrate, irradiated with the ions of Xe^+ with the energy of 300 keV (fluence $0.2 \div 2.0 \cdot 10^{16}$ cm^{-2}) at different temperatures. The results show that mutual mixing is determined by cascade mixing at temperatures below 160 °C and radiation-enhanced diffusion at a higher temperature. Mixing results in the formation of the phases Fe_3Si, FeSi and β-$FeSi_2$. The latter is the main phase at $T < 300$ °C. Radiation at 400 °C results in the formation of the α-$FeSi_2$ phase but the main phases at this temperature are Fe_3Si and FeSi.

The authors of [420] used the method of conversion electron Mössbauer spectroscopy for the investigation of the influence of the thickness of a zirconium substrate on ion mixing of two-layer materials Fe/Zr in bombardment with ions of Ar^+ with E = 200 keV with the dose in the range of 10^{14}–10^{17} cm^{-2}. It has been shown convincingly that the efficiency of mixing rapidly increases with increasing substrate thickness. Mixing resulted in the amorphisation of the two-layer film.

6.8. LONG-RANGE EFFECTS IN ION BOMBARDMENT

Conversion electron Mössbauer spectroscopy was used [338] to investigate the alloy $Fe_{70}Ni_{29.5}Mn_{0.5}$ in the initial state after quenching from 900 °C (100% of the FCC γ-phase) and after proton irradiation with the energy of 20 keV, dose $1.1 \cdot 10^{19}$ cm^{-2} (the calculated projected range of the ions was 120 nm). For the constant phase composition, it was found that the variation of the magnetic state of the alloy both on the irradiated and non-irradiated surfaces of a foil with a thickness of 10 μm. The changes were attributed to acceleration of the process of diffusion of the atoms of Fe and Ni because of the high degree of damage of the irradiated surface layer and the formation of concentration inhomogeneities in the microvolumes of the alloy at room temperature. The redistribution of the atoms of Fe and Ni at such a low temperature may, according to the authors, be stimulated by the presence in the alloy of implanted hydrogen characterised by high mobility allowing free diffusion to a large depth. The results show the variation of the Mössbauer spectrum of the surface of the specimen, irradiated for 48 hours, after the removal (by polishing on diamond paste) of a subsurface layer thicker than the projected path of the ions (Fig. 6.7). This is linked with the process of formation of regions with different nickel concentration.

The 'long-range' effects at a depth almost two orders of magnitude greater than the projected range of the ions, registered in the above study, were associated with the special nature of implanted ions, *i.e.* hydrogen ions, and mainly with the exclusively high mobility of hydrogen. Futher in this chapter, data are presented on the anomalously large depth of the ion irradiation effect of qualitatively different nature on the structure and properties of metal/alloys.

In [340, 379, 421, 422], the methods of transmission Mössbauer spectroscopy and x-ray diffraction analysis showed the non-thermal phase transitions (induced by ion bombardment) and atomic rearrangement in the entire volume of alloys $Fe_{69.5}Ni_{30.5}$, $Fe_{69}Ni_{31}$

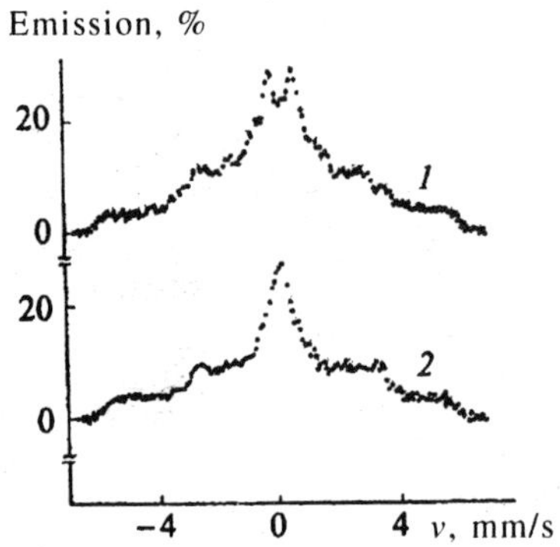

Fig. 6.7 CEMS spectra of $Fe_{70}Ni_{29.5}Mn_{0.5}$ alloy from the surface irradiated with H^+ [338], directly after polishing (*1*) and after holding for 48 h at room temperature (*2*).

and $Fe_{65}Al_{35}$. The heating of the targets was prevented by reducing the dose of irradiation with the ions of N^+ (or Ar^+) to $6 \cdot 10^{14} cm^{-2}$. In this case, the energy supplied to the specimens was simply insufficient to heat them to a temperature higher than 200 °C, even in the case of complete absence of heat removal by conductivity and radiation. The temperature of the targets during irradiation was controlled using low-melting reference specimens made of alloy POS-61 with a thickness of 30 μm (melting point 190 °C), which did not melt. The results show that the inverse (α→γ)-phase transition in the Fe–Ni alloys and the (atomic disorder) → (the long-range atomic order) transition in the $Fe_{65}Al_{35}$ alloy take place at $T < 200$°C and the radiation dose of $6 \cdot 10^{14}$ cm^{-2}, ovr the irradiation time of approximately 2 s (Fig. 6.8). The low values of the fluence, irradiation time and the relatively low temperature exclude from the number of possible reasons of the observed transformations the radiation-enhanced diffusion, chemical effect and static stresses from the implanted impurities (accumulate at high doses of irradiation).

The data obtained in [379, 421] by the method of nuclear gamma resonance (see Figs. 6.4 and 6.8) make it possible to separate quite reliably the processes associated with radiation-enhanced diffusion (formation of sublayers, enriched with Fe and Al, and spreading of the profile of N^+ in the subsurface zone with a thickness of 300÷400 nm at doses of 10^{17} cm^{-2} and larger) in the alloy $Fe_{65}Al_{35}$, and the process of atomic ordering of the entire volume of the foil with a thickness of 30 μm already taking place at the irradiation dose of $6 \cdot 10^{14}$ cm^{-2} during 2 s and in the absence of any significant heating ($T < 200$ °C).

In [422] the inverse phase (α→γ)-transition in the entire volume of the specimens of the $Fe_{69}Ni_{31}$ alloy with a thickness of 400 μm

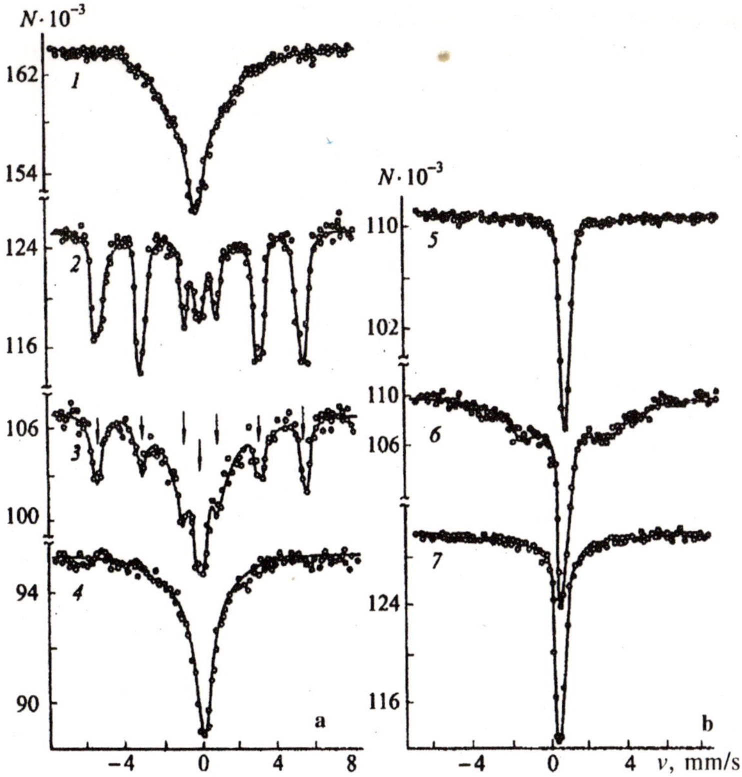

Fig. 6.8 Mössbauer transmission spectra [340, 379, 421] of alloys: *a*) $Fe_{69.5}Ni_{30.5}$: 1) FCC phase, produced by heating (680 °C, 1 h); 2) the state prior to irradiation after immersing the specimen in liquid nitrogen (BCC + residual FCC phase); 3) fluence $6 \cdot 10^{14}$cm^{-2}; 4) fluence $6 \cdot 10^{16}$cm^{-2}; *b*) $Fe_{65}Al_{35}$: 5) annealing at 650 °C, 6 h; 6) annealing + cold plastic deformation; 7) annealing + cold plastic deformation + irradiation, fluence $6 \cdot 10^{14}$cm^{-2}.

in irradiation with the ions of Ar (E = 20 keV, j = 300 μA/cm^2) was also observed in the cases in which the layers of Cu or Pt with a thickness of 0.1÷1.0 μm were deposited on the irradiated surface. In this case, the cascades of the atomic collisions were completely localised in the deposited films. The latter indicates that the transformations, enhanced by ion bombardment, are caused by post-cascade processes which are in all likelihood of the dynamic nature. The non-thermal phase transformations induced by ion bombardment at a depth of $10^3 \div 10^4$ R_p in [379, 423, 424] are linked with the propagation of post-cascade microshock waves of the type described in [342, 343]. In the normal conditions, these waves rapidly attenuate but in the metastable media they may become non-attenuating [423] because of the feed of energy liberated at the phase transformations, induced by these waves. In this case, the medium is transferred into a more thermodynamically stable state.

Thus, in addition to the destructive effect in the ion range zone, ion bombardment may have the exactly opposite effect on the state of the other anomalously deep subsurface layers of the materials. The radiation-enhanced phase transformations in metastable media are induced when some critical rate of feed of energy to matter is exceeded and they may self-propagate into the bulk of the substance with the velocities of the order or even higher than the velocity of sound [341, 342, 423, 424]. The transformations are characterised by exceptionally high efficiency because the energy, inducing transformation, is transferred only to the thin layer near the surface of the material and not to the entire transformed volume. In [422–424] an analogy is drawn between these phenomena and the phenomena of combustion and detonation.

As already mentioned, the investigated transformations are induced by very small fluences and take place at energy contributions insufficient for any significant heating of the specimens. The extent of transformation approaches 100% already after irradiation for several seconds. But if a spherical microshock wave breaks down into fragments of plane waves along different crystallographic directions, it will need to accumulate a significant dose to complete the transformation.

In [425], the method of transmission nuclear gamma resonance spectroscopy was used for study of foils (30 μm) of the $Fe_{69.5}Ni_{30.5}$ alloy in the initial α-(BCC) state (after holding at −196 °C) and after irradiation with continuous and pulse-repetitive beams of the ions of N^+ with the energy of 20 keV in different modes. The phase composition of the specimens was inspected by the method of x-ray diffraction analysis. In cases in which the density (or mean density) of the ion current was lower than 50 μA/cm^2, the (BCC → FCC) phase transition was not observed. However, in all caes when the mean density of current exceeded 50 μA/cm^2, the (BCC → FCC) phase transition was obseved in the entire depth of the irradiated foils of the above thickness. The processing of the Mössbauer spectra gave the functions of the density probability $P(H)$ of the distribution of effective magnetic fields on the nuclei of ^{57}Fe for the FCC phase of the $Fe_{69.5}Ni_{39.5}$ alloy, formed after annealing and different irradiation modes (Fig. 6.9). The functions $P(H)$ reflect the distributions of z-projections of the localised magnetic moments of the atoms of Fe and Ni in the iron–nickel alloy. In the first approximation, these functions could be

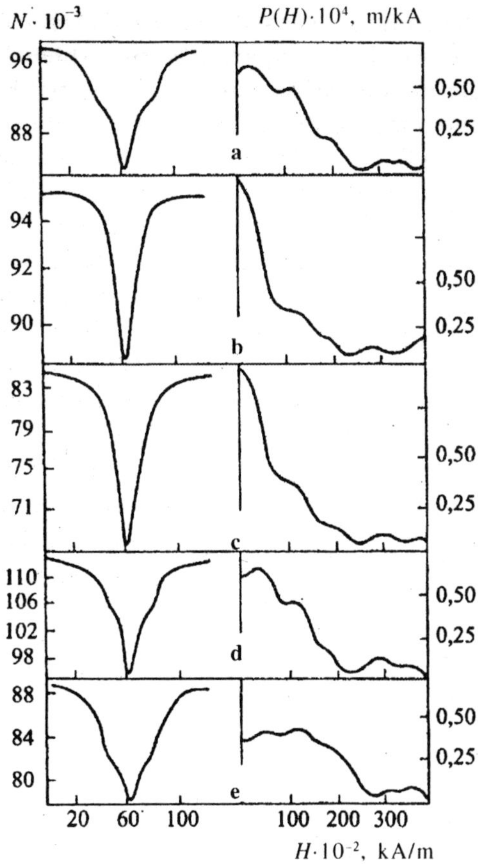

Fig. 6.9 Mössbauer transmission spectra [425] and the respetcive functions $P(H)$ of foils (30 μm) of $Fe_{69.5}Ni_{30.5}$ alloy for different treatment conditions (foils, prepared from sheets annealed for one hour at 650 °C and then held at –196 °C were irradiated with ions N^+): *a*) annealing at 650 °C, 1 h; *b*) irradiation in continuous regime, fluence $2 \cdot 10^{16}$ cm^{-2}, current density 50 μA·cm^{-2}; *c*) pulse–repetitive regime, fluence $5 \cdot 10^{16}$ cm^{-2}, ion current density 10 mA·cm^{-2}, frequency 5 Hz; *d*) $5 \cdot 10^{16}$cm^{-2}, 1 mA·cm^{-2}, 50 Hz; *e*) $5 \cdot 10^{17}$cm^{-2}, 1 mA·cm^{-2}, 50 Hz. In all cases, irradiation with ions N^+, E = 20 keV, pulse time 10^{-3} s.

approximated in all cases by the superposition of three Gaussian distributions with the centres $H_1 = (0 \div 0.5) \cdot 10^4$, $H_2 = (1.0 \div 1.2) \cdot 10^4$, $H_3 = (1.7 \div 1.9) \cdot 10^4$ kA/m, and the distribution intensities depend strongly on the treatment conditions.

It has been established that low fluences and high ion current densities are accompanied by the formation of highly non-equilibrium magnetic states (Fig. 6.9, curves b and c), which greatly differ from the state formed in annealing (curve a). A decrease of the current density in the pulse brings the thermal regime closer to conventional heating (curves d and e) and in the case of high fluences (curve e), it is apparently possible to reach a more equilibrium state in comparison with conventional heating (curve a), as a result of the specific effects associated with ion bombardment and mentioned in [425]. Some of the resultant magnetic states could be produced only in the application of superhigh pressures of several tens of kbar resulting in a decrease of the Curie point. Identical changes may evidently be associated with the redistribution of iron and nickel atoms in the alloy enhanced by ion irradiation.

In [426] the method of transmission Mössbauer spectroscopy on ^{57}Fe nuclei was used to investigate the ion irradiation-induced redistribution of atoms in foils (30 μm) of the $Fe_{93.75}Si_{6.25}$ alloy after preliminary maximum ordering by long-term stepwise annealing (800 °C – 1 h, 600 °C – 2 h, 500 °C – 4 h, 400 °C – 6 h, and 300 °C – 10 h). Irradiation was carried out with ions of N^+ with the energy of 20 keV and the ion current density of 50 μA/cm^2. The doses were 10^{17} and 10^{18} cm^{-2}.

The experimental and calculated profiles of the external peaks +1/2 → +3/2 and −1/2 → −3/2 of the Mössbauer spectra of the $Fe_{93.75}Si_{6.25}$ alloy, representing the superposition of the lines of the Lorentz form, are shown in Fig. 6.10a. The results of processing of these peaks (Fig. 6.10 b) indicate the increase of the degree of the short-range order in the solid solution after ion bombardment in comparison with the maximally ordered state, attained by thermal

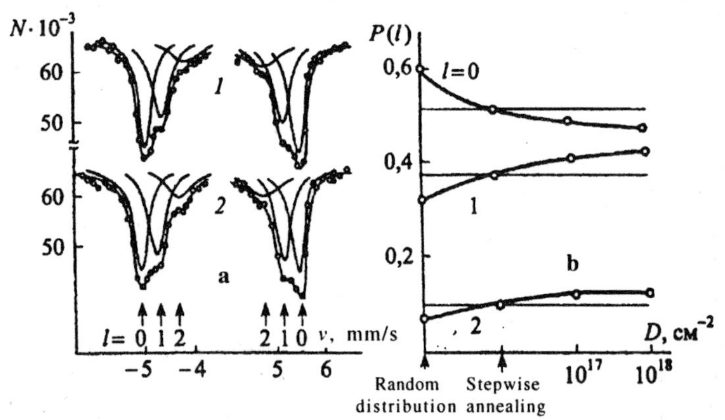

Fig. 6.10 Results of study of $Fe_{93.75}Si_{6.25}$ alloy [340, 379, 426]: *a*) experimental and calculated forms of external peaks +1/2 → 3/2 and −1/2 → −3/2 (*1* – stepwise annealing, *2* – stepwise annealing + irradiation with ions N^+, $D = 10^{17}$, 10^{18}cm^{-2}); *b*) probabilities *P*(*l*) of the Fe atom being surrounded by different numbers *l* of Si atoms in the alloy after different treatments, determined from the Mössbauer spectra and calculated for the case of random distribution of the atoms of Si and Fe in the crystal lattice sites.

treatment. The probabilities $P(l)$ of the presence of l atoms of silicon in the first coordination sphere of the iron atom, proportional to the areas S_l below the components of the external peaks, were determined by experiments. For the case of the random distribution of the atoms of Fe and Si in the alloy, the probabilities $P(l)$ were calculated from the relationship: $P(l) = C_z^l \, c^l(1-c)^{z-l}$, where C_z^l is the number of combinations of z in respect of l, $c = 0.0625$ is the silicon concentration in atomic fractions, z is the coordination number ($z = 8$).

Since measurements were taken in transmission geometry, the large variation of the probabilities $P(l)$ indicates that the process of ordering extends through the entire thickness of the specimen or, in any case, to the depth comparable with the thickness of the specimen. Additional ordering of the alloy is attributed to the excitation, during ion bombardment, of the low-temperature mobility of the atoms at temperatures below the thermal threshold of 'freezing' of diffusion. To explain the occurrence of low- temperature rearrangement of the atoms in the anomalously large depth, the authors of [423] and also of [340, 379] involve the representations of the generation of solitary shockwaves in the final stage of evolution of the cascades of atomic collisions.

In [427], investigations were carried out into the effect of ion bombardment (Ar^+, E = 20 keV, j = 100 μA/cm², $F = 5 \cdot 10^{16} \div 10^{18}$ cm^{-2}) and thermal annealing on the atomic and magnetic structure of $FePd_2Au$ alloy after 80% cold plastic deformation and quenching from 1200°C. Investigation by the Mössbauer effect and x-ray diffraction analysis showed that ion irradiation at 350 °C for 1.5–30 min results in the formation of a long-range atomic order in the disordered FCC matrix (a sublattice of iron atoms forms at the anomalously large depth, up to 20 μm; the projected ion range is ~13 nm). Ordering is accompanied by the ferromagnetic → asperomagnetic phase transition (T_{meas} = 77 K). Annealing at T = 350°C for up to 30 min in the absence of irradiation does not lead to any significant changes in the atomic and magnetic structure. The mobility of the atoms (the rate of formation of the ordered phase) in the course of irradiation at 350°C is approximately the same as in the course of annealing at 700°C.

The facts of occurrence of non-thermal phase transitions, induced by ion bombardment, and also of atomic intraphase rearrangement at temperatures considerably lower than the temperature of the start of intensive diffusion processes (diffusion 'de-freezing') at the anomalously large depth in the case of

$FePd_2Au$ and $Fe_{69}Ni_{31}$ alloys are confirmed by the data of field ion microscopy and measurements of the electrical resistance of the alloys directly during ion bombardment [423, 428, 429]. The data were obtained on the kinetics of the processes. The results show a very high rate of ordering, especially in the first seconds of irradiation, which, like the previously described facts, points to the difficulty of explaining the observed effects on the basis of the conventional representations of ion doping, formation of static stresses, thermal- and radiation-enhanced diffusion. The mechanism of these phenomena requires further investigation.

A review of studies of the Mössbauer spectroscopy of ion-doped metals and alloys was published in [430].

In [423, 427, 428] arguments are published in favour of the radiation-dynamic nature of non-thermal phase transitions induced by ion bombardment and determined by the propagation of microshock waves formed as a result of the evolution of dense cascades of atomic collisions. In metastable media, these solitary waves may become non-attenuating and sustained by the energy of phase transformations taking place at their front.

The radiation-dynamic effects can be studied by a unique source of sharply focused ion beams proposed in [431] and used in [432] for synthesis of a homogeneous $CoSi_2$ film on the silicon surface enabling almost any metallic ions to be produced with an energy of 20–50 keV and a maximum density of ion current up to 10 A/cm^2 (at an ion beam diameter of 100–2000 nm). The mode of single pulses with a minimum duration of 200 ns and a scanning pulse-repetitive regime with a frequency up to 5 kHz, resulting in uniform doping over an area of 160 × 160 μm, can be used. Since heating in the region of the 'prick' of the sharply focused beam does not exceed several degrees, this open the opportunities for the radiation-dynamic effects to be studied in the pure form, disregarding the role of the thermal effect.

6.9. CONCLUSIONS

Analysis of the content of the published studies indicates that the investigations by the Mössbauer effect have yielded exceptionally important information on processes taking place in metallic alloys during ion implantation (and also in subsequent heat treatment). Part of the data, obtained by the method of nuclear gamma resonance, confirms and supplements the results of investigations obtained using other methods.

In addition to this, completely new results have been obtained on the state of the surface and volume of ion-doped metals and alloys. This relates to the determination of the positions of implanted atoms and the structure of the resultant defects (including at extremely small quantities of implanted impurities), investigation of the stability and mobility of certain types of defects, identification of small amounts of phases, investigation of the parameters of the electrical and magnetic interaction of the implanted and matrix atoms, and also the detection and investigation of the long-range effects.

As shown by the results presented in the chapter, the processes taking place during ion bombardment in different layers over the depth of materials may greatly differ: 1) destructive processes (formation of defects) in the cascade zone, ion doping, formation of complexes from primary defects, impurity and matrix atoms; 2) amorphisation (at the minimum of thermal- and radiation-stimulated processes and the presence of chemical stimuli); 3) martensitic-type transformations at the depth several times greater than the projected range of the ions, stimulated by high static stresses from implanted impurities (and also in the case of the minimum role of diffusion); 4) subsequent migration of individual defects and of complexes of the defects, formation of vacancy voids (this increases the probability of formation of dislocation loops), the formation of concentration inhomogeneities, formation of segregations, pre-precipitates of the phases, *i.e.*, the occurrence of radiation-enhanced processes which together with heating, especially in implantation of mobile impurities, may greatly widen the zone of the effect; 5) heating of targets accompanying ion implantation[1] which becomes especially important in irradiation of the specimens, starting with the density of ion current (or mean density for pulsed-periodic beams) of several tens of $\mu A/cm^2$ (in the case of thin films rapid heating is most probable, especially in the case of inefficient heat removal); irradiation of the targets by ion beams with high mean power may result in changes of the properties of the targets at a large depth because of the reasons identical with those in the case of conventional heat treatment; 6) effects of initiation of phase transformations in layers with a thickness of up to $(10^3 \div 10^4)\ R_p$ and larger, not associated with heating and determined, in all likelihood, by dynamic post-cascade effects. In contrast to destructive effects

[1]The chapter does not include studies into the effect of powerful pulsed ion beams of the nanosecond range on the materials and no attention is given to the effect of formation of thermoelastic waves whose amplitude is high at a pulse time of $\tau < 10^{-7}$ s (see [337] for this problem).

in the ion range zone, these effects result in the transition of sub-millimetre surface layers of the initially metastable media to a more equilibrium state. This opens new possibilities of affecting the structure and properties of materials at reduced temperatures at which thermostimulated processes are absent.

The most complete pattern of the processes over the depth of the specimen may be obtained as a result of carrying out a selective (using a number of Mössbauer techniques) analysis of the surface layers of implanted materials. This analysis in the course of a single experiment is possible because of the unique design of the detector proposed in [346], combining the experiments on the transmission of γ-radiation, registration of secondary x-rays and internal conversion electrons. From the viewpoint of complex analysis of the surface and the bulk of matter, interesting results have also been obtained in [433].

Recently, extensive investigations have been carried out to develop methods of observing nuclei resonance scattering using synchrotron radiation. In particular, initial investigations were carried out to examine the surface of specimens in the sliding geometry which marked the beginning of Mössbauer spectroscopy of synchrotron radiation [434-437].

In [437] a new gamma-optical scheme of experiment was proposed for the differential (in respect of energy) grazing Mössbauer spectroscopy, based on simultaneous registration of: 1) gamma radiation, mirror-reflected by nuclei and electrons of the atoms; 2) conversion and Auger electrons; 3) characteristic x-rays, and 4) gamma radiation, resonance-scattered by the nuclei. Selective (in respect of depth) investigations of actual objects using synchrotron radiation were carried out for the first time. These investigations greatly widen the scope for studying the surface of solids, including that after the effect of radiation.

The number of investigations of ion-doped metals, using the Mössbauer effect, has been continuously increasing in the last years. This confirms the significant role of the method in investigating the local atomic and defect structure of crystals, and also the processes of formation of disperse phases, inhomogeneities of the composition, atom segregations, long-range effects, etc.

Appendix 1

Methods of Mössbauer spectra 'decoding'

A.1.1. DETERMINATION OF THE PARAMETERS OF PHYSICAL MODELS

As shown in chapters 1, 2, the Mössbauer spectra of alloys represent the sum of a relatively large number of elementary components. Knowing only the result of composition, *i.e.*, the experimental γ-resonance spectrum, it is often difficult to evaluate even approximately the values of the parameters of the individual components. It is easily understood that there is no purely empirical method of determination of the values of unknown parameters. At the same time, the investigated problem is solvable, and one of the main stages of solution of the problem is the modelling of the form of the resonance curve.

The base on the analytical model of the γ-resonance spectrum is the physical model constructed proceeding from the main representations of the theory of the Mössbauer effect and the theory of a real crystal.[1] It must take into account all the most important special features of the investigated objects: the presence of phases, their crystal structure, the type of atomic and magnetic ordering, dynamic state, etc.

It is quite impossible to provide unambiguous recommendations for the problems of modelling Mössbauer spectra. The most

[1]Depending on the level of modelling, the components of the vector of the parameters $\mathbf{x} = \{x_1, x_2,..., x_n\}$ have different physical meaning. In fact, parameters such as isomeric shift, the strength of the effective magnetic field on the nuclei of atoms, the width of the lines, etc, may be regarded either as independent or as functions of some specific characteristics of the solid. The level of modelling is determined by the aim of investigations, the level of development of theoretical representations and by a number of other factors. It is interesting to realise the principal possibility of modelling the resonance spectra from the first principles. However, it should be mentioned that any model has a meaning only if it is substantiated to a sufficient extent.

important aspects of the modelling of the Mössbauer spectra of the alloys, reflecting their specific features as objects of investigation, were investigated in chapter 2.

The modelling of the form of the spectrum requires accurate knowledge of the relationship of the parameters of the resonance curve with the characteristics of the solid. Only the values of the parameters are unknown. Statistical methods of determination of these parameters also require knowledge of the approximating function, i.e. the form of the γ-resonance spectrum. After writing the theoretical form of the resonance curve, the problem of determination of the unknown parameters can usually be solved by the methods of the theory of restoration of the parameters and relationships using empirical data. This second stage will be regarded as decoding of the Mössbauer spectrum.

It is fully natural to regard modelling as a more creative process than decoding, because from the viewpoint of the experimentator, the decoding procedure should represent in the ideal case a system of given (once for all) computing operations or computer programs. However, as shown by experience, the knowledge of the main principles of solution of such problems, the possibilities and special features of at least some of the frequently used methods of solution and special algorithms, and also programming elements, makes it possible to formulate more accurately the given problem of quantitative analysis, evaluate in advance the possibilities of obtaining one or the other information from the Mössbauer measurements and determine the possible sources of error.

The methods of decoding γ-resonance spectra

As already mentioned, the term 'decoding' has any meaning only if we know the effect of the variation of the physical characteristics (the vector of the parameters $\mathbf{x} = \{x_1, x_2, \dots x_n\}$) on the shape of the resonance curve, *i.e.* only when it is possible to describe analytically the form of the Mössbauer spectrum as a function of vector of $\mathbf{x}$. In this case, the problem of decoding is reduced to the determination of unknown parameters x_k (or x_1 and $f(x)$ in quasi-continuous description; in this case, the resonance curve can be transformed to the function of the distribution density of elementary components $f(x)$ with the minimum a priori information: see chapter 2, equation (2.9) and paragraph A1.3). When x_k or x_1 and $f(k)$ are determined, the theoretical form of the spectrum also becomes automatically unambiguously determined. In this case, if the physical model is adequate to the actual situation, then for the required values of x_1, x_2...x_n (respectively x_1 and $f(x)$) the theoretical curve of resonance absorption coincides with the experimental curve (within the limit of the error). The methods of finding unknown parameters will now be discussed.

The graphical method is one of the simplest methods used in the decomposition of the γ-resonance spectra into components and determination of the parameters of these components. The graphical method of analysis, undoubtedly, is useful in preliminary evaluation and obtaining qualitative data. At the same time, it is evident that the results of graphical analysis, because of their large error, cannot be used for the investigation of fairly fine phenomena and, in addition to this, the data obtained by graphical methods are highly subjective.

The currently available methods of evaluating the parameters are based on the application of objective goodness-of-fit, which are also criteria of the optimality of the solution. Historically, various criteria have been proposed. At present, preference is given to the method of maximum likelihood (ML), based on the application of the ML criterion. The errors of the values of the parameters, obtained by this method, are smaller in the majority of cases in comparison with any other methods.

The number of γ-quanta $I^*(v_k)$, obtained in experiments in a specific period of time, is determined by the relationships governing radioactive decay and is described by the Poisson distribution which in the case of sufficiently high $I^*(v_k)$ tends to the normal distribution (see, comment 1a in this Appendix). The conditions of the ML coincide in this case with the requirement of the minimum sum of the square of deviations of the theoretical and experimental intensities of the Mössbauer spectrum:

$$S(\mathbf{x}) = \sum_{k=1}^{R}\left[I(v_k\mathbf{x}) - I^*(v_k)\right]^2 / I^*(v_k) = \min, \tag{A1.1}$$

where R is the number of experimental points (it is assumed that R is considerably higher than n); $I(v_k, \mathbf{x})$ and $I^*(v_k)$ are the values of the theoretical and experimental intensity; $1/I^*(v_k)$ is the quantity reciprocal to the dispersion in the k-th point, which takes into account the different 'weight' of the individual points in the investigated sum because of a different statistical accuracy of the measurement of the values of $I^*(v_k)$.

Evidently, the sum (A1.1) may be regarded as a normal function of n variables s (x_1, x_2,..., x_n) or as the function of the vector of $\mathbf{x}$ ={x_1, x_2,..., x_n} in the n-dimensional space: $S(\mathbf{x}) \equiv S(x_1, x_2,..., x_n)$. Usually, $S(\mathbf{x})$ is referred to as the discrepancy function, and as the target function when we are discussing the search for the extremum $S(\mathbf{x})$. The essential condition of existence of the extremum of the function of many variables is the equality to zero of its partial derivatives:

$$\partial S(\mathbf{x})/\partial x_i = 0 \quad (i = 1, 2, ..., n). \tag{A1.2}$$

If the function $I(v_k, \mathbf{x})$ is linear in relation to all x_i, the solution of the system of linear equations, determined in this case by equality (A1.2), would represent the required vector $\mathbf{x}_{opt}$ corresponding to the unique extremum $S(\mathbf{x})$.

In many cases, including the problems of decoding Mössbauer spectra, the approximating function is nonlinear in relation to majority of x_i. In this case, the function may contain several extrema, with the global extremum being the 'deepest'. Since the system of equations (A1.2), obtained as a result of differentiation in respect of x_i, is nonlinear, the search for solutions of the system is quite complicated.

It would appear that the position of the global extremum $S(x_1, x_2,..., x_n)$ and, consequently, the value $\mathbf{x}_{opt}$ can be determined by direct calculation of the values of $S(\mathbf{x})$ in the nodes of the n-dimensional mesh. However, the number of calculations r of the target function rapidly increases with increasing n: $r = N^n$, where N is the number of nodes on the side of the n-dimensional mesh. Consequently, some problems cannot be solved in the realistically acceptable time even using the most powerful computers. These facts (the nonlinear form of the problem and cumbersome global examination of the function) have stimulated the development of special methods of determination of $\mathbf{x}_{opt}$.

The concept of one of these methods is the 'linearisation' of the function by expanding the function into a Taylor series with the accuracy to the linear terms in the vicinity of some point $\mathbf{x}^{(0)}$, being the initial approximation (to apply the methods, it is necessary to find the zero approximation by any way):

$$S(x_1, x_2, ..., x_n) \approx S(x_1^{(0)}, x_2^{(0)}, ..., x_n^{(0)}) + \sum_{i=1}^{n} \left[\frac{\partial S(\mathbf{x})}{\partial x_i}\right]_{\mathbf{x}=\mathbf{x}^{(0)},} \Delta x_i,$$
$$\Delta x_s = x_s - x_s^{(0)}. \tag{A1.3}$$

Subsequently, using the conditions

$$\frac{\partial S(x_1, x_2, \ldots, x_n)}{\partial \Delta x_p} = 0 \quad (p = 1, 2, \ldots, n), \tag{A1.4}$$

we obtain a system of n linear equations in relation to the n unknown quantities $\Delta x_p \equiv \Delta x_p^{(1)}$ (see comment 1b in the Appendix), and solving the system we obtain corrections for the values of $x_p^{(0)}$ and construct the vector $x_p^{(1)} = x_p^{(0)} + \lambda \Delta x_p^{(1)}$ $(0 < \lambda \leq 1)$; the methods of selection of λ were examined in, for example, [438]. The value of the vector of the parameters $\mathbf{x}^{(1)}$ is accepted as the initial approximation for the next iteration, and so on. It may be shown that if specific conditions are fulfilled, the sequence $\mathbf{x}^{(0)}$, $\mathbf{x}^{(1)}$, $\mathbf{x}^{(2)}$, ..., $\mathbf{x}^{(l)}$, ..., where l is the number of iteration, converges to x_{opt} [439].

The examined method is characterised by the high rate of convergence, is relatively simple and easy to program. The volume of the calculations and, consequently, the duration of computer solution of the problem increases in proportion to the square of the number of parameters n^2. Shortcomings of the method include the fact that in the presence of the dependence or a relatively strong correlation of the parameters, a degenerate matrix of the coefficient of the system of linear equations $||a_{ij}||$ may form. In this case, the system does not have the unique solution and the problem becomes unsolvable using the present method.

There are methods of bypassing situations like this, examined in special literature. However, in the absence of ready subprograms with the realisation of modified algorithms, it is often efficient to use different methods of solving the problem.

In addition to the 'linearisation' method, it is also possible to use the methods of finding the minimum $S(\mathbf{x})$, based on comparing the values of the target function for different values of the vector $\mathbf{x}$, which change in accordance with the search procedure, and on the investigation of certain local properties of the multidimensional surfaces $S(\mathbf{x})$.

The simplest and at the same time sufficiently effective method of finding the extremum of functions of many variables is the Gauss–Zeidel method [440] based on coordinate-by-coordinate search for partial minimum values of the target function with the preset the value of the step. Search for each consecutive parameter starts from the point reached in the process of search on the previous coordinate. After exhausting all directions, the cycle is repeated. If the movement along all coordinates with the given step becomes impossible (the approximate value of the minimum with the given step is found), the step is 'broken up', usually two–four times.[1] The process of search continues until reaching the required accuracy in respect of x_k (i.e., until the search step becomes smaller than the required accuracy). If necessary, other criteria may also be used.

The Gauss–Zeidel method is most efficient when the surfaces of the level of the target function are close to hyperspheres, and it is not suitable if there is a relationship between the parameters x_j (the surfaces of the level form 'gullies'). The volume of calculations increases with increasing r more rapidly than in the method of linearisation and is proportional to n^3. However, the comparison of the efficiency of individual methods for each specific class of problems can be carried out most reliably by direct evaluation of their efficiency.

Figure A1.1 shows the scheme, illustrating the operation of the Gauss–Zeidel method when $n = 2$.

A suitable algorithm named 'the pattern search' method was proposed by R. Hooke and T. Jeeves [441]. The logical structure of the method makes it possible to hold in the process of finding the extremum on the edge of the sharp 'ridge' ('gully' in searching for a minimum). The algorithm of the method was described in [208]. Figure A1.2 shows the nature of movement along the gully using the pattern strategy. According to D. Wilde [208], the pattern

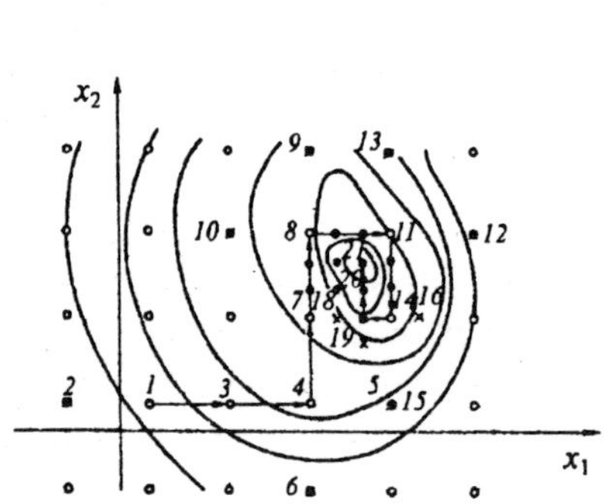

Fig. A1.1 The work of the simplest scheme of the Gauss–Zeidel methods, two-dimensional variant.

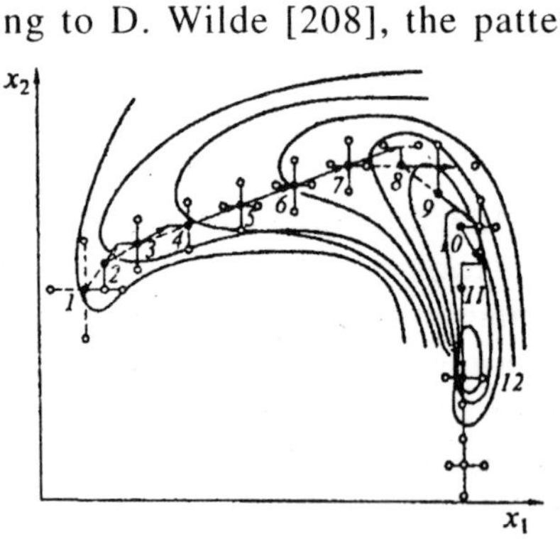

Fig. A1.2 (right) Tactics of tracking the ridge in the pattern search method.

---

[1]It may be shown that if the surfaces of the levels are represented by hyperspheres, the highest rate of convergence is attained by dividing a step e times (e is the base of natural logarithms).

search method is most suitable for solving the problems associated with the approximation of the experimental dependences by the method of least squares. R. Hooke and T. Jeeves showed by experiments that, in this case, the duration of search for the extremum increases in proportion to n, *i.e.* it is considerably slower in comparison with the Gauss–Zeidel and linearisation methods.

A sufficiently large number of different methods of finding the extremum and modifications of these methods are available at the present time. However, hardly any of these methods have any major advantages in comparison with other methods. The efficiency of a specific method for different classes of problems may differ. In addition to this, the criteria of the evaluation of applicability and efficiency of algorithms may also differ: the duration of search for the extremum, the extent of global examination of the function, the accuracy of the resultant solution, etc. In addition to the named methods, it is also important to mention the methods of parallel tangents, and also a group of gradient methods, based on the motion in the direction of the gradient. A special position is occupied by the random search methods, in which the direction of movement is determined using the probability scheme.

A.1.2. EVALUATION OF ERRORS AND CORRELATION COEFFICIENTS

After finding the minimum of the target function $S(\mathbf{x})$, and, correspondingly, the vector of the optimum estimates of the determined parameters $\mathbf{x}_{\text{opt}} \equiv \tilde{\mathbf{x}} = \{\tilde{x}_1, \tilde{x}_2, ..., \tilde{x}_n\}$, it is interesting to evaluate the degree of reliability of the resultant values of $\tilde{\mathbf{x}}_k$. It appears that the information on the magnitude of the errors (standard deviations of the parameter $\sigma_{\tilde{x}_i}$) and the correlation coefficients $\rho_{\tilde{x}_i,\tilde{x}_j}$ is found in the structural matrix $\|a_{ij}\|$ (see the comment 1b in Appendix 1), calculated at $\mathbf{x} = \tilde{\mathbf{x}}$. In the method of linearisation, similar matrices are formed in each iteration. When using other methods, the matrix $\|a_{ij}\|_{\mathbf{x}=\tilde{\mathbf{x}}}$ should be calculated using a special procedure. According to [65, 442], the confidence interval for the estimates $\tilde{x}_i$ may be expressed in the following form:

$$\left(\tilde{x}_i - t_{a,r}\sigma_{\tilde{x}_i},\quad \tilde{x}_i + t_{a,r}\sigma_{\tilde{x}_i}\right), \tag{A1.5}$$

where $t_{a,r}$ is determined from the confidence probability α (usually, it is assumed that α = 0.95) and the number of degrees of freedom

$r = R-n$ from the tables of the Student distribution law. The estimates $\sigma_{\tilde{x}_i}$ are determined from the relationship:

$$\sigma_{\tilde{x}_i} = \sqrt{a^{ii} \frac{Q(\tilde{x})}{R-n}}, \tag{A1.6}$$

where a^{ii} is the diagonal element of the matrix $\|a^{ij}\|_{\mathbf{x}=\tilde{\mathbf{x}}}$, reciprocal in relation to $\|a_{ij}\|_{\mathbf{x}=\tilde{\mathbf{x}}}$; $Q(\tilde{\mathbf{x}})$ is the residual sum of the squares $\left(Q(\tilde{\mathbf{x}}) \approx S(\tilde{\mathbf{x}})\tilde{x}_1\right)$; R is the number of experimental points, n is the number of parameters. The final result may consequently be presented in the form $x_i = \tilde{x}_i \pm \Delta x_i \left(\Delta x_i = t_{a,r}\sigma_{\tilde{x}_i}\right)$.

Correlation coefficients $\rho_{\tilde{x}_i,\tilde{x}_j}$ are calculated from the equation

$\rho_{\tilde{x}_i,\tilde{x}_j} = a^{ij} \dfrac{Q(\tilde{x})}{\sigma_{\tilde{x}_i}\sigma_{\tilde{x}_j}(R-n)}$, where a^{ij} are the corresponding elements of the reciprocal matrix.

A.1.3. METHODS OF RESTORATION OF THE FUNCTIONS OF DENSITY OF THE DISTRIBUTION OF PARAMETERS OF HYPERFINE INTERACTION

The problem of decoding the Mössbauer spectra is a typical example of the restoration of the values of parameters and relationships on the basis of experimental data. As already mentioned, in formulation and solution of the problems of this type it is initially necessary to utilise the a priori information to the maximum extent. The information from the theory of the methods, symmetry, etc. makes it possible in a number of cases to model the expected type of dependences and the gamma-resonance spectrum as a whole with the accuracy to the unknown parameters. The values of these parameters can be obtained using conventional statistical methods reducing the problem of finding the minimum of the nonlinear functional.

However, in some cases, it is difficult to propose a substantiated model of the gamma-resonance spectrum or give preference to one or the other model. The verification of the existing hypotheses is especially difficult when the non-resolved or poorly-resolved spectra are analysed. It is therefore interesting to apply mathematical methods of restoration of the functions of the density of distribution of the individual parameters using the experimental data.

Restoration problems of this type (in this case, subintegrand functions are restored) relate to the class of ill-posed problems. Specially developed methods of solution of such problems [443–445], see also appendix 2, are now available.

In the studies, concerned with the analysis of gamma-resonance spectra, the determination of the density of distribution of the HFS parameters is often carried out using two different approaches: the B. Window method [446], and the method proposed by J. Hesse and A. Rubartsch [447], and modifications of these methods. To improve the accuracy of terminology, it should be mentioned that the Window method corresponds to the method of quasi-solutions [445]. Hesse and Rubartsch used the regularisation method [443, 444].

The justification of the applicability of these methods to the analysis of gamma-resonance spectra is reduced to justification of representing the resonance curve in the form of a linear superposition of continuously distributed 'subspectra', differing in the value of some parameter of hyperfine interaction:

$$\varepsilon(v) = \int_{x_{min}}^{x_{max}} p(x) f(x,v) dx, \qquad (A1.7)$$

where v is Doppler velocity; $\varepsilon(v)$ is the absorption effect; $f(x, v)$ is the form of the elementary subspectrum as a function of parameter x; $p(x)$ is probability density.

According to physical relationships, this description is justified. In fact, the resultant spectrum, especially in complicated systems, characterises the presence of a large number of non-equivalent positions of resonant nuclei in the crystal. The comparatively strong effect of the first coordination sphere which differs for the atoms with different environment, determines the existence of several centres of distribution of the HFI (hyperfine interaction) parameters. The weaker effect of each subsequent coordination sphere results in the 'diffusion' of these centres. Consequently, the quasi-continuous distribution of the HFI parameters is obtained. In pure metals and in compounds, the density $p(x)$ may be represented in the form of a single or of a sum of several δ-functions. It should be mentioned that the relaxation effects made play a significant role in the formation of the hyperfine structure of the spectrum.

The Window method

The method is based on the expansion of $p(x)$ into a series[1] and is examined with reference to the case in which the HFI parameter is the effective magnetic field H. The existence of a continuous distribution of the magnetic hyperfine fields in some section $[0, H_{max}]$ is assumed. The experimental set of the values of $\varepsilon^*(v_j)$ is approximated by the expression:

$$\varepsilon(v_j) = \int_0^{H_{max}} p(H) f(H, v_j) dH, \tag{A1.8}$$

where $f(H, v_j)$ is the Zeeman sextet of the lines with the splitting, proportional to H. The density of distribution of the hyperfine fields is represented, according to Window [446, 447], by the expansion:

$$p(H) = \sum_{i=1}^{n} a_i \left[\cos\left(i\pi H / H_{max}\right) - (-1)^i\right]. \tag{A1.9}$$

The coefficients a_i are determined using the method of least squares. Assuming that:

$$S = \sum_{j=1}^{R} \left[\varepsilon^*(v_j) - \varepsilon(v_j)\right]^2 = \min, \tag{A1.10}$$

i.e. $\partial S / \partial a_k = 0 \quad (k = 1, 2, \ldots, n),$ (A1.11)

where R is the number of experimental points, and solving the resultant system of n linear algebraic equations (A1.11) in relation to a_k, the required vector $\{a_1, a_2, \ldots a_n\}$ and the form of $p(H)$ are determined.[2]

The Hesse–Rubartsch method

In this method, the integral (A1.8) is replaced by a sum with some small step $H_i = ih$; $i = 0, 1, 2, \ldots, l$. Consequently,

$$\varepsilon(v_j) = \sum_{i=0}^{l} p(H_i) F(H_i, v_j) h, \tag{A1.12}$$

---

[1]In terms of a set of orthogonal functions, for example, into a Fourier series. It should be mentioned that the Window expansion yields almost the same type of $p(x)$ as the Fourier cosine expansion.

[2]If the normalisation condition $\int_0^{H_{max}} p(H) dH = 1$ is not taken into account in the equation (A1.8), the function $p(H)$ must be normalised.

or

$$\varepsilon(v_j)=\sum_{i=0}^{l} p_i F(H_i,v_j), \qquad \text{(A1.13)}$$

where $F(H_i, v_j)$, is, as in the Window method, the sextet of the lines; in this case, the density $p(H)$ degenerates into a discrete set of probabilities $p_k = p(H_k)h$ (k = 0, 1, 2,..., l). The concept of regularisation is then employed. The values of p_k are determined from the condition:

$$\frac{\partial}{\partial p_k}\left\{\sum_{j=1}^{R}\left[\varepsilon^*(v_j)-\varepsilon(v_j)\right]^2+\gamma\sum_{i=2}^{l-1}\left[\left(p_{i-1}-2p_i+p_{i+1}\right)^2\right]\right\}=0, \qquad \text{(A1.14)}$$

where γ is the regularisation parameter (the parameter of smoothness of the solution). The resultant expression determines the system of l + 1 linear equations in relation to l + 1 unknown values p_k. It should be mentioned that at γ = 0 the condition (A.14) corresponds to the usual requirement of the minimum of the sum of the squares of deviations of the experimental and approximating spectra but, as indicated by practical experience, the solution of the system (A1.14) gives in this case the 'non-physical' set of the values p_k (lower smoothness of the solution, large negative spikes, etc). By selecting the value of γ it is possible to obtain the acceptable type of distribution of p_k and describe the form of magnetically split spectrum with sufficient accuracy. The reference point for the selection of γ in the Hesse–Rubartsch method was the normalised residual sum of the squares ($S_N \approx 1$) [447].

In [447] the authors carried out a detailed comparative analysis of the described methods. It has been shown that for the appropriate selection of parameter γ and the number of the expansion members n, both methods provide close solutions. The practical realisation of the Hesse–Rubartsch method requires shorter computing time but its application is associated with the solution of systems of the equations of a considerably higher order in comparison with the Window method, which, in some cases, makes it necessary to restrict examination to processing only part of the Mössbauer spectrum. Evidently, the problem of the selection of the suitable method for solving one or the other problem must be solved individually in every case. Taking into account the important applied value of the methods of restoration of the density functions for a wide range of the problems in gamma-resonance spectroscopy, and also the fact that analysis of certain problems, required for the direct practical application of this method, is not available in the literature, special attention will be given to the main moments associated with the restoration of the function of density of the HFI parameters on the example of the quasi-solution method [139, 448].

Procedure for the determination of p(x) by the quasi-solution method

The given description of the form of the resonance curve (A1.8) assumes that the mathematical expectation of the experimental value of the absorption effects (see equation (1.3)) $\varepsilon^*(v)$ is equal to $\varepsilon(v)$, *i.e.* $\varepsilon(v_j) = \varepsilon^*(v_j) - \Delta_j / I(\infty)$, where Δ_j is the error at the j-th point, measured in pulses, $I(\infty)$ is the count at infinity (the estimate $I(\infty)$ may be obtained with sufficient accuracy by averaging the intensity values of the 'wings' of the spectrum). In this case, equation (A1.8) may be presented in the new form:

$$\varepsilon^*(v_j) - \frac{\Delta_j}{I(\infty)} = \int_{x_{\min}}^{x_{\max}} p(x) f(x, v_j)\, dx. \tag{A1.15}$$

To determine $p(x)$, it is convenient to use the Fourier cosine expansion in the section $[x_{\min}, x_{\max}]$:

$$p_n(x) = \sum_{k=0}^{n} a_k \cos\frac{k\pi x'}{L}, \tag{A1.16}$$

where $L = x_{\max} - x_{\min}$, $x' = x - x_{\min}$, n is the number of the highest harmonic. The requirement of the minimum of the residual sum of the squares taking into account the difference of the statistical weights of the individual points $\sum_{j=1}^{R} \omega_j \Delta_j^2 = \min$, in the substitution of the equations (A1.15) and (A1.16), is presented in the following form:

$$S = I^2(\infty) \sum_{j=1}^{R} \frac{1}{I^*(v_j)} \left[\varepsilon^*(v_j) - \sum_{k=0}^{n} a_k \int_0^L \cos\frac{k\pi x'}{L} f(x', v_j)\, dx' \right]^2 = \min, \tag{A1.17}$$

where $\omega_j = 1/I^*(v_j)$ is the statistical weight of the j-th point of the spectrum. The differentiation of equation (A1.17) in respect of a_k

($\partial S/\partial a_k = 0$; $k = 0, 1, \ldots, n$) gives the system of linear equations, and the coefficients of these equations are given by the relationships:

$$a_{ij} = \sum_{k=1}^{R} \frac{1}{I^*(v_k)} \int_0^L \cos\frac{i\pi x'}{L} f(x', v_k)\,dx' \int_0^L \cos\frac{j\pi x'}{L} f(x', v_k)\,dx', \qquad \text{(A1.18)}$$

$$\beta_i = \sum_{k=1}^{R} \frac{\varepsilon^*(v_k)}{I^*(v_k)} \int_0^L \cos\frac{i\pi x'}{L} f(x', v_k)\,dx'. \qquad \text{(A1.19)}$$

The matrix of the coefficients of the left-hand parts is symmetric. The solution of the system has the form of the vector $\{a_0, a_1, \ldots, a_n\}$ which unambiguously determines the type of $p_n(x)$. It should be mentioned that since the subintegrand functions for the harmonics of a higher order are rapidly oscillating, special attention must be given to the accuracy of calculating the integrals. The relationships (A1.18) and (A1.19) make it therefore possible to obtain the approximation for $p(x)$ and calculate the type of approximating spectrum (A1.7).

One of the main problems associated with the application of the considered restoration methods is the selection of the number of terms of the expansion in the quasi-solution method and of the regularisation parameter (smoothness of the solution) in the regularisation method.

The author of this book and his colleagues proposed a method of selecting the value n on the basis of a statistical criterion[1] χ^2 [139, 448]. Since the mathematical expectation of χ^2 is equal to $R - m$ (m is the number of coefficients a_k in the expansion (A1.16), equal to $n+1$), and the mean quadratic (standard) deviation is $\sqrt{2(R-m)}$, then in the case of the adequate description of the experimental spectrum the value of the quadratic functional S should be close to $R-m$ at a sufficiently narrow confidence interval of the order of $\pm\sqrt{2(R-m)}$. The results of experimental investigation of the behaviour of the residual sum of the squares (A.17) as the

---

[1]The number of pulses I_j, registered in the j-th channel of the spectrometer, is governed by the Poisson statistics. In this case, since I_j is sufficiently high ($10^5 \div 10^6$ pulses), the distribution of the errors Δ_j is Gaussian, and the distribution of the quantity $S = \sum_{j=1}^{R} \omega_j \Delta_j^2$ is described by function χ^2 [65]).

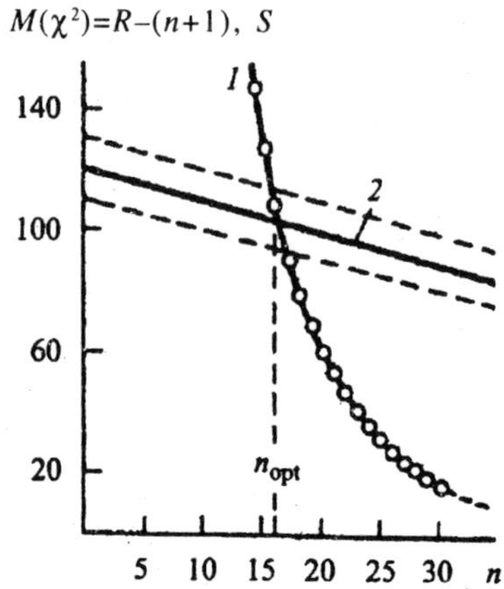

Fig. A1.3 Determination of the optimum number of terms of expansion of the series (A1.16): 1) normalised residual sum of squares (S); 2) mathematical expectation of the random quantity χ^2.

function of the number of the highest harmonic n are generalised in Fig. A1.3 which explains the method of selection of the optimum value $n = n_{opt}$. Whilst at $n < n_{opt}$ the description of the form of the experimental spectrum is insufficiently accurate (S rapidly increases with decreasing n), at $n > n_{opt}$, 'over-description' (excessively detailed description, i.e. the repetition of the insignificant components of the form of statistical nature) occurs. The latter is accompanied by the formation of non-physical oscillations in $p(x)$, in particular, the number and amplitude of negative spikes increases. Thus, the optimum number of harmonics should result in the value of S in the immediate vicinity of $R-m$. A similar criterion may also be used in the selection of the smoothness coefficient in the regularisation method.

Because of the restrictions imposed on the number of terms of the expansion of function $p(x)$ into a series, it is necessary to examine the problem of the resolving power of the method of restoration with reference to the problems of Mössbauer spectroscopy. Since the number of the expansion terms cannot exceed some optimum value, it is evident that the limiting resolution should be represented by the value comparable with the period of the highest harmonics in the expansion (A1.16). For example, at the number of the terms of expansion equal to n, the application of the quasi-solution method makes it possible to mark and reproduce bursts of density $p(x)$ whose width in any case is not lower than $(0.3 + 0.5) \cdot 2L/n$ (L is the expansion interval). The bursts of the smaller width will be smoothed and widened as a result of restoration In this sense, it is evident that the method of regularisation is also characterised by a limited resolving power. In this case, we are concerned with the restrictions associated with the statistical nature of the information, and not with the special features of the restoration methods. Evidently, one of the methods of increasing the resolution power is the increase of the number of experimental points in the investigated range of distribution of the HFS parameters.

Form of the function f(x, v_j)
In a general case, the Mössbauer spectrum may be represented by the superposition of elementary components with different resonant energy, having the Lorentzian form in the limit of the thin absorber:

$$f(x,v_j)=\frac{1}{1+4/\Gamma^2(x-v_j)^2}. \qquad \text{(A1.20)}$$

Using (A1.20) as the subintegrand function in the expression (A1.7) gives the density of distribution of the shifts (the positions of the centres of gravity) of the elementary components of the resonance curve. In this case, it is necessary to know independently only two parameters: $I(\infty)$ and Γ. The value of $I(\infty)$ can be determined with sufficient accuracy by the level of the 'wings' of the experimental spectrum (if necessary, corrections can be made for the geometry). With a decrease of the partial thickness of the elementary component the value of Γ should tend to the double natural width of the line ($2\Gamma_0$); in the actual conditions, it is slightly higher. Experience with the processing of gamma-resonance spectra shows that the variation of Γ in a relatively wide range (0.19÷0.25 mm/s) has only a slight effect on the form of resultant density function $p(x)$. It should be mentioned that the function of the type (A1.20) can be used in the presence of hyperfine interactions of any nature. The resonance curve is transformed in this case to the function of the density of positions of the centres of gravity of the Lorentzians (i.e. to the so-called 'true spectrum'). The obtained true spectrum is characterised by a higher resolution power in comparison with the initial spectrum.

In the analysis of the distribution of the values of the quadrupole and magnetic dipole splitting, the form of the elementary components is respectively specified by one of the following methods:

$$f(x,v_j)=\frac{1}{1+4/\Gamma^2(v_j-c-x/2)^2}+\frac{1}{1+4/\Gamma^2(v_j-c+x/2)^2}, \qquad \text{(A1.21)}$$

or

$$f\left(x,v_j\right)=\sum_{i=1}^{6}\frac{A_i}{1+4/\Gamma^2\left(v_j-c-B_ix\right)^2}. \qquad \text{(A1.22)}$$

It may be seen that in these cases it is also necessary to specify the position of the centre of gravity c which may be easily calculated as the mean value of the coordinate of the spectrum $c=\sum_{j=1}^{R}v_j\varepsilon^*(v_j)\Big/\sum_{j=1}^{R}\varepsilon^*(v_j)$. In processing of magnetically split spectra, it is also necessary to have the vector of intensity of the sextet lines (if the different directions of magnetisation in the domains have the same probability on the average over the crystal, then **A** = {3, 2, 1, 1, 2, 3}). Vector **B** is fully determined by the nuclear constants (**B** = {0.5; 0.289; 0.079; –0.079; –0.289; –0.5}).

It is evident that in the substitution of equations (A1.21) or (A1.22) into equation (A1.7) the symmetric spectrum is synthesised, whereas the experimental absorption curves are not always symmetric, even in the case in which the dominant factor is one of the considered types of hyperfine interaction (quadrupole or magnetic dipole). The non-symmetric form of the gamma-resonance spectrum may be determined by the superposition of the electrical monopole spectrum on quadrupole and magnetic dipole, or by the presence of combined magnetic and electrical quadrupole interactions. In cases in which there is a correlation (in the limiting case, a functional dependence) between the HFI parameters, asymmetry may be taken into account by setting the corresponding relations, for example, between δ and H (in equation (A1.22) it is assumed that $c = c(x)$). In this case, the form of the function $f(x, v_j)$ changes respectively. The non-symmetric form of spectrum of a different nature may also be taken into account when writing the form of $f(x, v_j)$.

Examples of solutions of problems

The proposed procedure of restoration of the functions of the distribution density of the HFS parameters by the quasi-solution method in the form of a program in Fortran-IV language was used by the author of the book and colleagues for analysis of the distribution of H_{ef} in magnetically ordered Fe-Mn alloys with 20 and 41 at.% of Mn [139] and for analysis of the HFS of the ferromagnetic alloys FeAl with 5, 10, 15 and 20 at.% of aluminium

[448].

The convergence of the algorithm was verified by synthesis of the model magnetically split spectrum:

$$I(v_j) = I(\infty)\left[1 - \int_0^{H_{max}} p(x) f(x, v_j) dx\right] \quad (A1.23)$$

without the application of statistics, where x has the meaning of the current value of H_{ef}; the density of distribution of the fields is represented by the Gaussian function:

$$p(x) = \frac{1}{\sqrt{2\pi}\sigma} \exp\left[-\frac{(x-\bar{x})^2}{2\sigma^2}\right], \quad (A1.24)$$

where $\langle H_{ef} \rangle \equiv \bar{x} = 34$ kOe, σ = 2.3 kOe (H_{max} = 86 kOe). Function $f(x, v_j)$ was defined by the relationship (A1.22); Γ = 0.29 mm/s; $\mathbf{A}$ = {3, 2, 1, 1, 2, 3}, and the shift of the centre of gravity of the spectrum and quadrupole splitting were assumed to be equal to 0. Figure A1.4 shows the form of the synthesised spectrum, and Fig. A1.5 the type of function $p(x)$ and its restored values $p_{11}(x)$ and $p_{21}(x)$, obtained for the number of terms of expansion of the series (A1.16) n = 11 and n = 21. It may be seen that with increase of n the quality of approximation improves. At n = 21, the calculation spectrum almost completely coincides with the spectrum synthesised

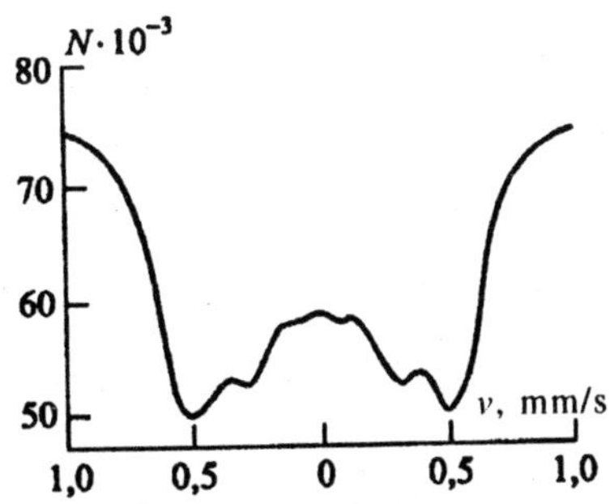

Fig. A1.4. Synthesised spectrum of Fe+40 at.% Mn alloy.

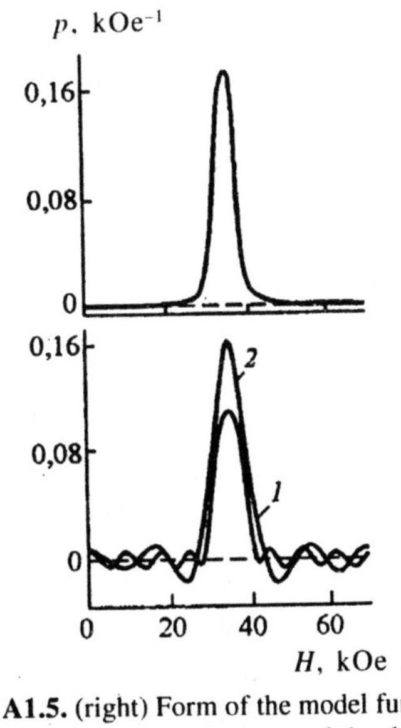

Fig. A1.5. (right) Form of the model function $p(H)$ (a) and restored values of the density $p_{11}(H)$ (1) and $p_{21}(H)$ (2) (b).

using equation (A1.23). (It should be mentioned that in all cases, the number of points R should be sufficiently large to ensure that the ratio $L/(R-1)$ is considerably smaller than the period of the highest harmonics $2L/n$). The results of application of the programme for the processing of Mössbauer spectra of the two-phase alloys Fe +21 at.% Mn (the structure contains the FCC γ-phase and HCP ε-phase) and the single-phase Fe + 40 at.% Mn alloy (100% of the γ-phase) are shown in Fig.A1.6. The characteristic feature of the two-phase alloys with 21 at.% of manganese is the presence of two distribution centres H_{ef}. It has been established that additional alloying with the third component may result in the formation of additional peaks $p(x)$. This is attributed to the formation of concentration inhomogeneities.

The external peaks of the Mössbauer spectrum of the non-ordered ferromagnetic alloys with 10 at.% Al and the resultant functions of the density of distribution of the centres of gravity of the continuously distributed components $p(x)$ are presented in Fig. A1.7. The external peaks of the spectrum are almost completely unresolved. Nevertheless, the corresponding density functions show distinctive maxima, corresponding to the atoms of Fe^{57} with $l = 0$, 1 and 2 atoms of aluminium in the nearest neighbourhood. The calculation of the densities $p(x)$ [204, 448] for the alloys with 5, 10, 15, 20 at.% Al (see chapter 5) gave the values of the HFS parameters without any modelling presentations regarding the atomic–crystalline structure and the nature of hyperfine interactions in FeAl alloys.

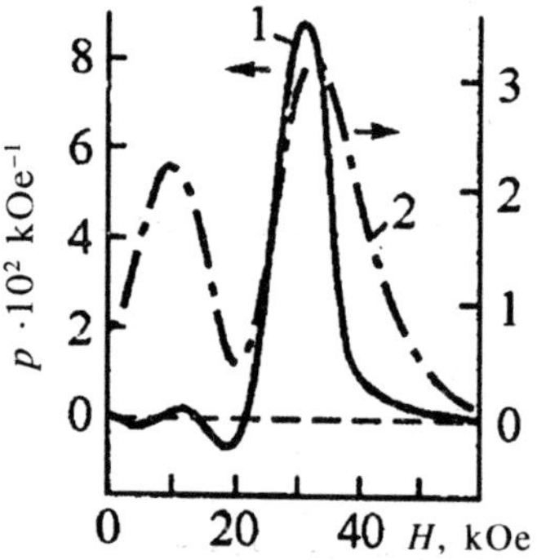

Fig. A1.6 Densities of distribution of effective magnetic fields: 1) Fe + 40 at.% Mn; 2) Fe + 21 at.% Mn (temperature –150 °C).

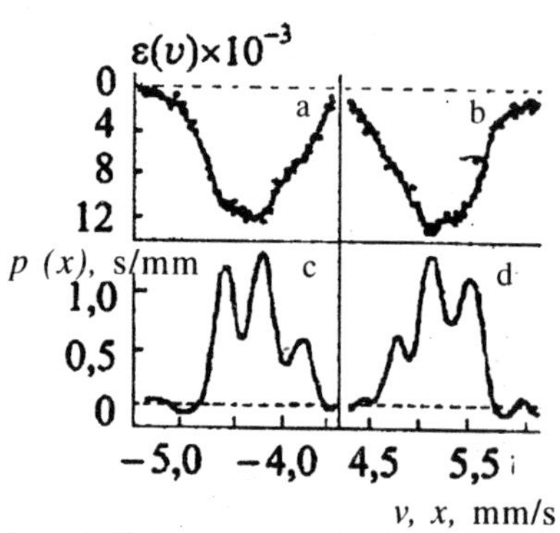

Fig. A1.7 External peaks (+1/2 → +3/2, –1/2 → –3/2) of the Mössbauer spectrum of the Fe + 10 at.% Al alloy (*a*, *b*) and the functions of the density of centres of gravity of the elementary components (*c*, *d*) restored from them.

The probability of the presence of a different number of aluminium atoms in the first coordination sphere of iron, calculated from the ratio of the areas of the peaks $p(x)$ is in good agreement with the results of calculation of these probabilities assuming the random distribution of the atoms in the alloys. The application of the method of restoration of the density functions is sufficiently promising for 'model-less' evaluation of the parameters of the HFS and obtaining information, required for constructing the physical models.

At present, the methods of restoration of the density functions are used for analysis of the alloys with a complicated magnetic structure, including the cases of coexistence of different types of magnetic order, investigation of magnetic ordering in thin films, etc.

A.1.4. THE METHOD OF DIFFERENCE SPECTRA

This method is based on relatively simple assumptions, with the main assumption being that the material subjected to some external effect may be regarded as a set of ensembles of nuclei (atoms) in the *changed* and *unchanged* states in relation to the initial state. For a sufficiently thin absorber, the areas of the coresponding components of the Mössbauer spectrum are proportional to the number of nuclei (atoms) in one and the other states. Therefore, if the spectrum of the initial specimen with the 'proper weight' (see below) is subtracted from the spectrum of the specimen subjected to the effect, one obtains a subspectrum corresponding to the atoms in a changed state. It is desirable to ensure that the conditions of registering of the compared specimens of Mössbauer spectra are as close as possible. To eliminate the existing differences, the spectra may be adjusted by reno-rmalisation.

The form of the difference spectrum is determined as follows:

$$Y(v_i) = Y_1^n(v_i) - kY_0(v_i), \tag{A1.25}$$

where $Y_0\ (v_i) = I_0(\infty) - I_0(v_i)$, $Y_1^n(v_i) = I_0(\infty) - I_1^n(v_i)$ are the absorbed intensities for the initial specimen and the specimen subjected to the effect (the normalised intensity $I_1^n(v_i)$ is determined from the experimentally measured dependence $I_1(v_i)$ as a result of the renormalisation procedure, described below); v_i is the Doppler velocity; k is the fraction of the atoms with the properties, identical with the properties of the ensemble of ^{57}Fe atoms in the initial state.

The renormalisation procedure of the spectrum includes the following operations. Initially, the spectrum $I_1(v_i)$ is transformed to the spectrum:

$$I_1^{'}(v_i) = I_1(v_i)\frac{I_0(\infty)}{I_1(\infty)} \qquad \text{(A1.26)}$$

(the values $I_1(\infty)$ and $I_0(\infty)$ are determined as the average values of the ordinate on the 'wings' of the spectra). Consequently, the equality of the intensities in the absence of resonance is achieved: $I'_1(\infty) = I_0(\infty)$. This is followed by renormalisation, removing the difference of the integral intensities (areas) of the peaks. For this purpose, it is necessary to calculate the areas below the resonance curves:

$$S_0 = \sum_{i=1}^{R} Y_0(v_i) \quad \text{and} \quad S_1 = \sum_{i=1}^{R} Y_1'(v_i), \qquad \text{(A1.27)}$$

where $Y_1'(v_i) = I_0(\infty) - I_1'(v_i)$, R is the number of experimental points. This is followed by the determination of the normalised (both in respect of the intensity 'at infinity' and in the area) value of $Y_1^n(v_i)$:

$$Y_1^n(v_i) = \frac{S_0}{S_1} Y_1^{'}(v_i), \qquad \text{(A1.28)}$$

and, finally, by the procedure of calculation of the difference spectrum (A1.25).

The values of the ordinates of the normalised and difference spectra may now be obtained from the relationships:

$$I_1^n(v_i) = I_0(\infty) - Y_1^n(v_i), \qquad \text{(A1.29)}$$

$$I(v_i) = I_0(\infty) - Y(v_i). \qquad \text{(A1.30)}$$

Coefficient k is determined from the experiment on the basis of the following considerations. If it is assumed that the spectrum of a specimen subjected to some effect, is the superposition of the subspectra from the atoms in the *changed* and *unchanged* states, then in the case of the correctly selected value of k, the intensity of the unchanged component in the composition of the difference spectrum (A1.25) should become zero. A further increase of k should result in the formation of non-physical negative values of intensity $Y(v_i)$ of the difference spectrum. Additionally, when using the method of difference spectra, it is necessary to take into account the fact that in the case of the correctly selected value k the difference spectrum should have the maximally simple form. In addition to this, it is necessary, if possible, to estimate independently the fraction of the atoms $(1 - k)$ with the changed properties.

The method of difference spectra may be very useful in the analysis of non-resolved and poorly resolved spectra. It is important that in contrast to the fitting methods, this method does not require any additional (*a priori*) information which may prove to be inaccurate or erroneous.

Comment 1a. In the unchanged conditions (geometry, resonance conditions, etc), the probability of detection of I γ-quanta during the measurement time τ in the Mössbauer experiment is determined by the laws of radioactive decay and is given by the Poisson distribution:

$$p(I) = \left(I_0^I / I!\right) e^{-I_0}, \tag{A1.31}$$

where I_0 is the mathematical expectation of the random quantity I. As indicated by equation (A1.31), the number of γ-quanta, observed in a concrete measurement, is only approximately equal to I_0, although the probability of finding the values of I, close to I_0, is maximum. It is well-known that at $I_0 \to \infty$, the distribution $p(I)$ tends to the normal distribution with the mathematical expectation $M(I)$ and the dispersion $D(I)$ equal to I_0:

$$p(I) = \left(2\pi I_0\right)^{-1/2} \exp\left(-\frac{(I - I_0)^2}{2I_0}\right). \tag{A1.32}$$

This relationship is a sufficiently good approximation already at $I_0 = 10^2 \div 10^3$. In the Mössbauer measurements, I_0 is usually equal to $10^5 \div 10^6$. At high values of I_0, the distribution (A1.31) may be regarded as quasi-continuous, and $p(I)$ may be regarded as the probability density. It should be mentioned that at $I_0 \to \infty$, the relative scatter of the values of I tends to zero:

$$\sigma / I_0 = \sqrt{D(I)} / I_0 \doteq 1/\sqrt{I_0} \to 0 .$$

Comment 1b. In the explicit form, the coefficients of the system of linear equations (A1.4) taking into account (A1.1) are given by the relationships:

$$a_{ij} = \sum_{k=1}^{R} \frac{1}{I^*(v_k)} \left[\frac{\partial I(v_k, \mathbf{x})}{\partial x_i} \frac{\partial I(v_k, \mathbf{x})}{\partial x_j} \right], \tag{A1.33}$$

$$b_i = \sum_{k=1}^{R} \frac{I^*(v_k) - I(v_k, \mathbf{x})}{I^*(v_k)} \left[\frac{\partial I(v_k, \mathbf{x})}{\partial x} \right], \tag{A1.34}$$

where a_{ij} are the coefficients of the left-hand part, b_i is the vector of the right-hand part. The matrix $||a_{ij}||$ is called the structural or constructional matrix.

Appendix 2

Linear ill-posed problems. Inverse problems of spectroscopy

In 1929, the French mathematician Jacques Hadamard introduced the concept of the well-posed problem, formulating three conditions which the problem must satisfy. The problem is referred to as well-posed according to J. Hadamard if its solution: 1) does exist; 2) is stable, and 3) is unique. The problems, which do not satisfy at least one of these conditions, are referred to as ill-posed problems. The most interesting and content-rich case forms when the second condition, *i.e.* the condition of stability of the solution, is violated.

For a long time, many scientists have believed that the ill-posed problems have no right to exist and, in particular, cannot be found in physical and technical applications because they have no physical meaning.

Nevertheless, the classical formulation of certain problems of natural science resulted in many cases in mathematical problems with unstable solutions or, in accordance with the given definition, in ill-posed problems. The disruption of stability indicates that the change of the solution may be arbitrarily large at any however small variation of the initial data of the problem.

There are physical processes that are unstable because of their nature at some critical values of the parameters, characterising these processes. For example, when an aircraft passes through the sound barrier, there is a sharp change in the field of velocities, the pressure field and other parameters of the gas medium, interacting with the aircraft (and also the derivatives of these parameters in respect of the spatial coordinates and time). Attention will however not be given to the problems of this type but to a relatively wide range of the problems, associated with the interpretation of the results of observations (restoration of the dependences on the basis of empirical data) which in its classical formulation are unstable and, consequently, ill-posed, but they can be reduced to stable in the case of accurate (correct) formulation, taking the nature of the problem adequately into account.

The most significant contribution to the theory of linear ill-posed problems has been provided by the Russian mathematicians

A.N. Tikhonov and V.K. Ivanov. In 1966, their studies were awarded the Lenin Prize. It is also important to note the significant role of the studies published by M.M. Lavrent'ev as well as R. Lattês, J.-L. Lions and B.L. Phillips at the stage of emergence of this trend of research.

Prior to examining the wide range of problems, associated with the formulation and solution of the ill-posed problems, the terminology used in this case will be discussed briefly. In particular, attention will be given to the words and word combinations which are either directly associated with the examined subject or have contact points and entire regions of intersection with theory and practice of solution of ill-posed problems.

The natural science, technical and also physical–mathematical literature contains terms and word combinations such as: unstable problems (solutions), problems of interpretation of the results ofobservations (restoration of relationships from empirical data), non-classical methods of restoration of dependences, inverse problems of spectroscopy (or problems of restoration of the true spectrum), gravimetry, plasma physics, etc, problems of restoration of images, two-level image recognition schemes, the problems of filtration of noise, signals, numerical differentiation; convolution (convolution equation), integral Volterra equation, Fredholm equation of the first kind, the response function (instrument function, the kernel of the integral equation), the linear response operator, regularisation of the solution, regularising operator, smoothing functional, stabilising functional, regularisation parameter (the parameter of smoothness of the solution), the function and the generalised discrepancy method.

A.2.1. LINEAR ILL-POSED PROBLEMS

In the most general form, the linear ill-posed problem may be formulated as follows [443, 444, 449]:

$$Az = u, \tag{A2.1}$$

where $z \in Z$ and $u \in U$ are the elements of the metric spaces, A is the linear operator (the linear response operator). Usually z and u are vectors or functions (*i.e.* the vectors in the functional space). Definitions of these concepts are not presented here, they can be found in, for example [450]. It is important to note the two characteristic properties of a linear operator: 1) $A(z_1+z_2) = Az_1+$

Az_2, 2) $Aaz_1 = aAz_1$. In further considerations, the specific examples, described by the operator equation (A2.1), will be discussed.

In the problems of interpretation of the results of observations it is assumed that A is the operator known from the theory (theory), u are the results of observations (experiment), and z is the unknown (required) dependence or the discrete series of numbers, restored as a result of solving equation (A2.1) (solution), representing this dependence. Since the results of measurements (observations) are always burdened with errors, the classical formulation of the problem of restoration of the dependences on the basis of the empirical data is usually formulated in the form:

$$Az = u^*, \tag{A2.2}$$

where the designation u^* stresses the fact that the components of the vector u, obtained as a result of measurements, contain errors. If the operator A is a matrix operator, equation (A2.2) may be presented in the form:

$$\left(\sum_j\right) A_{ij} z_j = u_i^* \quad (i = 1, 2, ..., n; j = 1, 2, ..., m), \tag{A2.3}$$

where A_{ij} are the elements of the matrix A. The sign of the sum is often omitted and it is understood that summation is carried out in respect of the repeating index (the Einstein rule). It is assumed that $n \geq m$, *i.e.* in the examined case, equation (A2.1) has the form of the system of n algebraic equations with m unknown quantities. For example, at $m = n = 2$ there is a system of two linear equations with two unknown quantities:

$$\begin{aligned} A_{11}z_1 + A_{12}z_2 &= u_1^*, \\ A_{21}z_1 + A_{22}z_2 &= u_2^*. \end{aligned} \tag{A2.4}$$

Prior to transferring to analysis of the stability of some typical problems, it is necessary to present a simple and convincing example, which has no relationship with natural sciences, but makes it possible to demonstrate the consequences of the instability of solutions.

It is assumed that we have purchased the first time A_{11} units of a specific product with the price z_1, and A_{12} units of another product at the price z_2 and paid, for both products, $u_1 = 6.59$ monetary units, and another time required A_{21} and A_{22} units of the same products of the same prices (z_1 and z_2) and paid for the purchase $u_2 = 3.24$ monetary units. It is assumed that the $A_{11} = 1.001$; $A_{12} = 2.059$; $A_{21} = 0.494$ and $A_{22} = 1.010$ (in the units of measurement of the amount of these products). We shall now try to restore the prices of both goods, if the quantity u_1 was determined with the error of +0.02 (the erroneous number is indicated by the asterisk):

$$
\begin{aligned}
&1.001z_1 + 2.059z_2 = 6.61^*,\\
&0.494z_1 + 1.010z_2 = 3.24.
\end{aligned}
\tag{A2.5}
$$

Solving the system of equations gives $z_1 = -0.81$ and $z_2 = 3.60$ units (this is complete nonsense). It should be seen that if we would have substituted the value 6.59 to the right part of the first equation, we would have obtained, as can be easily verified by substitution, the accurate values $z_1 = 2.49$ and $z_2 = 1.99$.

Thus, in practice we have confirmed the instability of the system of equations (A2.4) because a small (in any case, expressed in percent) variation of the initial data ($|\delta u_1/u_1| \approx 0.3\%$) results in a very large change (distortion) of the solution. In subsequent stages, it is assumed that δu and δz are the mean quadratic values of the errors for the set of the numbers u_i and z_i.

In a general case, if the matrix operator is defined and we have a system of equations of type (A2.3), then at $m = n$ and det $\neq 0$ (main determinant is not equal to zero) there is a unique solution of this equation: $z = A^{-1} u$, where A^{-1} is the matrix, reciprocal in relation to A. Since the solution z depends in a continuous manner on u, from $\delta u \to 0$ we obtain that $\delta z \to 0$ (it is always possible to select a sufficiently small $\varepsilon > 0$ which is such that at $\delta u < \varepsilon$ the condition $\delta z < \delta$ is fulfilled, where δ is some however small number defined in advance) and, consequently, the problem is theoretically stable. However, in practice, the problem may prove to be unstable, as indicated by the discussed example. In particular, the errors of the right-hand part δu may determine variations of the solution δz which are so large that the absolute value of the error exceeds the absolute value of the 'true' solution (corresponding to $\delta u = 0$) or, in the language of problems of filtration of noise, the noise will exceed the restored signal (although $\delta u \to 0$ gives $\delta z \to 0$ and, formally, the problem is stable).

The problems of interpretation of the results of observation lead quite often to integral Volterra or Fredholm equations of the first kind [443, 444]. The representation of the operator equation (A2.1) in the form of the integral Fredholm equation of the first kind has the following form:

$$Az = \int_a^b A(x,t)\,z(t)dt = u(x), \quad c \le x \le d, \tag{A2.6}$$

where $A(x, t)$ is the function available from theory (the kernel of the integral equation), $u(x)$ is the measured function, which depends on parameter x, $z(t)$ is the required function which should be restored on the basis of the results of observation (measurements). Evidently, if we had the absolutely accurate theory at our disposal, and also the a priori known function $z(t)$ and the undistorted, *i.e.* containing no errors, results of observation $u(x)$, the relationship (A2.6) would convert to identity.

When the kernel of equation (A2.6) depends only on the difference of the variables x and t, the equation transforms to the convolution equation:

$$Az \equiv \int_a^b A(x-t)\,z(t)dt = u(x), \quad c \le x \le d, \tag{A2.7}$$

representing the partial case of the Fredholm equation.

Let us show that the problem of determination (restoration) of $z(t)$, formulated in the form of equation (A2.6) is unstable, *i.e.* ill-posed. For this purpose, following [443, 444], we introduce the auxiliary function $z'(t) = z(t) + a\sin\omega t$. Since the operator A is linear, then $Az' = Az + aA\sin\omega t$. Let ω tend to infinity ($\omega \to \infty$). Evidently, in this case, for any point of the section $[c, d]$:

$$\delta u(x) = aA\sin\omega t = a\int_a^b A(x,t)\sin\omega t dt \to 0, \quad c \le x \le d, \tag{A2.8}$$

since the subintegrand function represents the rapidly oscillating function, modulated by the comparatively slowly changing (relatively smooth) function, describing the real physical or other processes (Fig. A2.1). However, regardless of the fact that at $\omega \to \infty$ the maximum deviation of the right-hand part on $[c,d]$ is $|\delta u|_{max} \to 0$, selecting $a > N$, where N is any, however high number, specified in advance, we obtain that $|\delta z|_{max} > N$.

Consequently, as a result of measurements, because of the

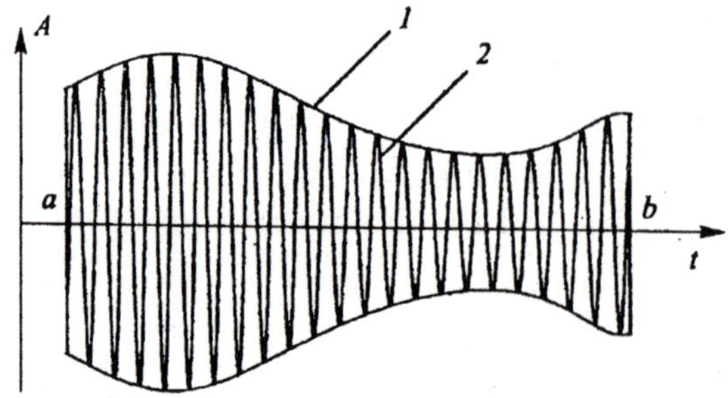

Fig. A2.1 Function $A(x, t)$ (1) and product $A(x, t)\sin\omega t$ (2) ($x \in [c, d]$).

presence of errors (distortions, noise), we may obtain the function $u^*(x) = u(x) + \delta u(x)$ which differs however little from $u(x)$ which, however, may correspond to any, however large distortion of the solution $z(t)$. This means that the formulated problem of restoration $z(t)$ is ill-posed because of the disruption of the condition of stability of the solution.

In the problems of interpretation of the results of observations, the measured function is usually represented by the discrete set of the values $u^*(x_i)$ ($i = 1, 2, \ldots, n$) burdened with the errors, *i.e.* we obtain the system of n intergral equations from which we can either restore the characterising $z(t)$ discrete set of the values $z(t_j)$ ($j = 1, 2, \ldots, m$) or obtain approximation for $z(t)$, for example in the form of the expansion into a series in the section $[a, b]$ (see paragraph 6 of this appendix).

Since the integral in equation (A2.6) may be approximately represented by the integral sum, we can write the following approximate equality:

$$\sum_{j=1}^{m} A(x_i, t_j) z(t_j) \Delta t_j \approx u^*(x_i), \tag{A2.9}$$

which becomes more accurate with an increase of the number of divisions of the section $[a, b]$ at $(\Delta t_j)_{max} \to 0$. Denoting $A(x_i, t_j)\,\Delta t_j = A_{ij}$, $z(t_j) = z_j$ (in many cases, it is convenient to use the uniform step $\Delta t_j = \tau$), we obtain the relationship

$$A_{ij} z_j \approx u^*_i \quad (i = 1, 2, \ldots, n; \quad j = 1, 2, \ldots, m) \tag{A2.10}$$

which differs from (A2.3) by the fact that it is approximate. In other words, the problems of the solution of the Volterra and Fredholm equations of the first kind may be approximately reduced to solving the system of linear algebraic equations which is often used in practice.

A2.2. INVERSE PROBLEMS OF SPECTROSCOPY

The range of the linear ill-posed problems, described by equation (A2.3), is extremely large. It includes the problems of linear algebra (which may show an instability because of the presence of rounding errors), differentiation of inaccurately specified functions, finding of the solutions of integral equations of the first kind, summation of the series with the approximately specified coefficients, etc.

The linear ill-posed problems form in the solution of greatly differing natural science and applied problems. These are problems of restoration of the distributions of the potential, plasma diagnostics (for example, obtaining the distributions of the electronic temperature of the plasma by the x-ray method), restoration of images, in particular, problems of computer tomography. The problems of this type relate to the group of the so-called inverse problems.

The most illustrative and easy-to-grasp problems of this type are the inverse problems of spectroscopy: optical, x-ray, Mössbauer (nuclear gamma-resonance), etc. In this group, we may also include the problems of determination of the distribution of flows of microparticles (having, like electromagnetic quanta, wave properties), by the energies, for example, the problem of determination (restoration) of energy-dependent differential nuclear sections of absorption of neutrons by the known integral neutron spectra.

The inverse problem of spectroscopy is formulated as the problem of the restoration of the initial spectrum $z(t)$ from the instrument spectrum $u^*(x)$ (obtained in the spectrometer). Here, t and x have the meaning of frequency for the frequency or of energy for the energy spectrum. Every spectrometer is characterised by the so-called instrument function (or the response function of the spectrometer) $A(x, t)$ which is determined by the physical principles of operation and the design of the spectrometer and is usually known with the degree of accuracy required for measurements. For the optical spectrometer, the instrument function is usually represented by the Gaussian:

$$A(x,t)=A_0\exp\left[-(x-t)^2/2\sigma^2\right], \tag{A2.11}$$

where A_0 characterises the amplitude of the signal, and σ is the resolution power of the spectrometer (the minimum variation of frequency of energy, recognised by the spectrometer). As the value of σ decreases, the resolution power increases. The relationship between $A(x, t)$, $z(t)$ and $u^*(x)$ in the classical formulation of the problem of restoration of $z(t)$ on the basis of the empirical data is determined by the relationship:

$$Az \equiv \int_a^b A(x,t)\,z(t)dt = u^*(x), \quad c \le x \le d, \tag{A2.12}$$

which differs from (A2.6) by the fact that in this case the results of observations $u^*(x)$ burdened with errors are used.

If the true spectrum $z(t)$ is represented by a very narrow bell-shaped line and $t = t_0$ with the 'width' at half the height tending to zero (in Fig. A2.2a this line is represented by the vertical stroke), the instrument spectrum $u(x)$ (without taking errors into account) represents the instrument function (A2.11) with the centre at the point $x = t_0$, *i.e.* $u(x) = A(x, t_0) = A_0 \exp[-(x - t_0)^2/2\sigma^2]$ (Fig. A2.2b). Actually, the infinitely narrow true spectrum may be represented by the Dirac δ-function: $z(t) = \delta(t - t_0)$, whilst $u(x) = A(x, t_0)$ follows directly from (A2.6) taking one of the properties of the δ-function into account [450].

Figure A2.2a also shows the true spectrum $z(t)$, representing the

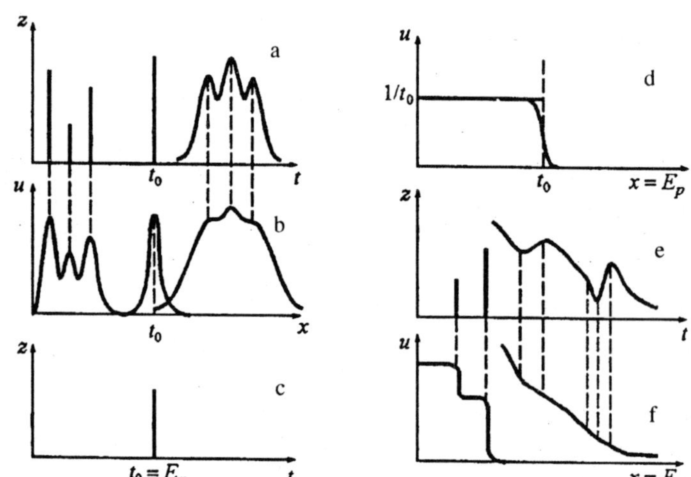

Fig. A2.2 Examples of true (*a*, *c*, *e*) and instrument (*b*, *d*, *f*) spectra with response functions of the type (A2.11) and (A2.13).

sum of three δ-functions, and the true spectrum $z(t)$ representing the continuous distribution with three maxima. The respective instrument spectra $u(x)$ are presented in Fig. A2.2b.

Another example will be discussed. The functions of energy distribution of reactor neutrons are determined using a hydrogen-containing detector (or a detector on recoil protons), usually a stilbene crystal. It is well-known that a neutron with the energy of t_0 in collision with a proton (having practically the same mass as the neutron) may transfer to the proton any energy x (with equal probability from 0 to t_0 (at head-on collision, the neutron is stopped and the proton acquires the energy $x = t_0$). The response function of the spectrometer with the detector on the recoil protons (or the instrument function) will in this case evidently have the form of a 'step':

$$A(x,t) = \begin{cases} A_0 / t_0, & x \le t_0, \\ 0, & x > t_0, \end{cases} \tag{A2.13}$$

where $1/t_0$ is the fraction of the recoil protons per unit energy interval. In the experiment, the value of A_0 is proportional to the neutron flux, the efficiency of formation of recoil protons and the exposure time. It is assumed that the neutron is subjected to no more than one collision in the stilbene crystal, *i.e.* the probability of a collision is small.

For monoenergetic (monochromatic) neutrons, with energy $E_n = t_0$, the instrument spectrum coincides with the instrument function (A2.13); in the actual case, the step is slightly smoothed out (Fig. A2.2, c, d). It may also contain other special features because of the presence of secondary effects which, however, does not change the essence of the problem. Figures A2.2d and f show examples of the differential ('true') and integral (instrument) spectrum in the discrete and continuous energy distribution of the neutrons.

A.2.3. INCONSISTENCY OF SOME 'OBVIOUS' APPROACHES TO SOLVING INVERSE PROBLEMS

For the majority of problems, operator A is either given in the matrix form or may be reduced to this form. It is therefore natural to try to determine the values of z_j simply by solving the system of linear algebraic equations (A2.10), similar to (A2.3). Actually, the values of u_i^* are available from observations, and the elements. A_{ij}

of the operator A can be determined (calculated) from theory.

However, as shown in paragraph A.21 of the present appendix, illustrating this on a specific example, the system of linear algebraic equations of type (A2.3) may prove to be unstable in practice, *i.e.* in the presence of relatively small errors in the initial data there may be a catastrophic distortion of the solution. If it is attempted to solve the system (A2.10) and restore the true spectrum $z(t)$ (indicated by the continuous line with three maxima in Fig. A2.2a) on the basis of the values of $u^*(x_j)$ of the instrument spectrum measured with errors, we obtain the absolutely 'non-physical' solution of the type shown in Fig. A2.3b since the problem is unstable.

This may easily be shown in practice by synthesising (defining) model spectra $z(t)$ in the form of any functions which we may like and, subsequently, calculating the integral (A2.6) (with the instrument function in the form (A2.11) or (A2.13)) at points x_j (representing in detail function $u(x)$, i.e. at least in 50÷100 points). If this is followed by adding, to the resultant values of $u(x_j)$, random or quasi-random errors (this may be carried out using the program of the 'generator of random numbers') and solving subsequently the system of equations (A2.10) then, even if the random error δu_i does not exceed $0.001 u_{i\,\max}$ in any of the points, the solution will almost certainly have the form similar to the solution shown in A2.3b.

In fact, setting the exact equality between Az (where the operator A is exactly known from theory) and vector u^*, distorted by errors, we obtain unavoidably a distorted solution which, by analogy, should be denoted by z^*. Thus, to restore the vector z, adequately representing the 'true' solution using the distorted

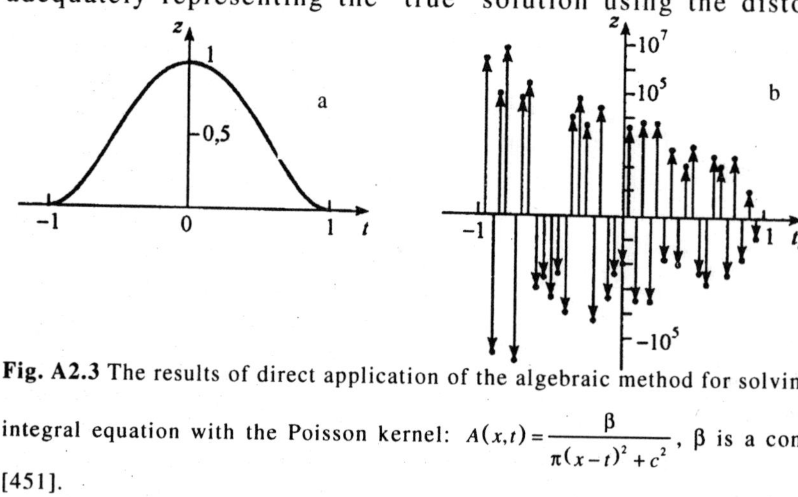

Fig. A2.3 The results of direct application of the algebraic method for solving the integral equation with the Poisson kernel: $A(x,t) = \dfrac{\beta}{\pi(x-t)^2 + c^2}$, β is a constant [451].

experimental data, it is necessary to think of a more adequate formulation of the problem and write its mathematical form, taking into account the fact that in this case the left and right parts of the equation (A2.2), *i.e.* theory and experiment, should correspond to each other only with a certain degree of approximation, determined by the error of the initial data.

Let us ιry to change the approach to a certain extent. The exact equality may be written by deducting from the right-hand part the error of the i-th measurement $\rho_i \equiv \delta u_i$:

$$\sum_{j=1}^{m} A_{ij} z_j = u_i^* - \rho_i. \tag{A2.14}$$

With this, it is accepted that the left part differs from the right part because of the presence of errors. Unfortunately, we do not know the values of these random errors ρ_i. However, it is possible to estimate z_j by the criterion (and method) of least squares (MLS).

It is necessary to ensure that the sum of the squares of deviations. $S = \rho^2(Az,u) = \sum_{i=1}^{n}(1/\sigma_i^2)\rho_i^2$ in the minimum, *i.e.*

$$S = \rho^2(Az,u) = \sum_{i=1}^{n} \frac{1}{\sigma_i^2} \left(\sum_{j=1}^{m} A_{ij} z_j - u_i^* \right)^2 = \min. \tag{A2.15}$$

The multiplier $1/\sigma_i^2$ takes into account the different statistical weight, *i.e.* different accuracy, of individual measurements.

The necessary condition of existence of the extremum is the condition of the equality of partial derivatives to zero:

$$\frac{\partial S}{\partial z_k} = 0 \quad (k = 1, 2, \ldots, m). \tag{A2.16}$$

After differentiation, the following equation is obtained:

$$A^* \, Az = A^* u^* \tag{A2.17}$$

which at $m = n$, as may easily be seen, does not differ at all from (A2.3): the multiplication of the left and right parts by the same matrix operator, as can easily be confirmed, for $m = n$ does not change the solution (here A^* is the matrix transposed in relation to A, in which the lines are replaced by the columns and vice versa). The minimum of the functional (A2.15) is obtained at $Az = u^*$, *i.e.* at $\rho^2(Az, u) = 0$. The requirement of the minimum of the functional in this case corresponds to the requirement of the exact equality

(A2.3). If m slightly exceeds n, the form of solution is qualitatively the same and may also be illustrated by Fig. A2.3. In a general case, further accumulation of statistical information, *i.e.* increase of n, does not guarantee the stability of the solution. However, in some cases, the premediated decrease of m results in the increase of the stability of the solution. These methods are sometimes referred to as the methods of intuitive regularisation (the term 'regularisation' is introduced in paragraph A2.6). They are used in practice for increasing the stability and improving the type of solution. However, the application of each specific algorithm is justified only if it is strictly substantiated.

Thus, the method of least squares on its own cannot be the basis for the solution of the investigated class of the problems.

A.2.4. THE NON-CLASSICAL APPROACH TO THE RESTORATION OF RELATIONSHIPS

We shall analyse examples of the description of the relationships, which may be capable of explaining the resultant situation. If it is assumed that $A_{ij} = x_i^j$, $u_i^* = y_i^*$, and $z_j = a_{j+1}$ (where x_i^j is the '*eks*' i-th to the power j), then the problem $Az = u^*$ may be regarded as identical with the problem of description of the experimental dependences by the polynomials of the k-th degree: $y(x) = a_0 + a_1x + a_2x^2 + \dots +a_kx^k$ $(y_i = a_0 + a_1x_i + a_2x_i^2 + \dots + a_kx_i^k)$[1].

Figure A2.4 shows the results of some hypothetical experiment (points y_i^*) from $n = 9$ measurements together with the indication of the values of the errors. There are also variants of the description of these data by the polynomials of different degree k by the method of least squares, and also the curve, drawn intuitively (by hand).

It may be seen that the description by the parabola (Fig. A2.4a) represents 'with insufficient accuracy' the special features of the experimental curve, although, according to the criterion of the least squares method, the depicted parabola gives the best description of the experiment (of the entire set of parabolas). The polynomial of the fourth degree (Fig. A2.4b) transfers 'externally with sufficient accuracy' the special features of the experimental dependence. The curve, representing the graph of the polynomial of the $(n-1)$-th

[1]The situation when a theoretical curve is presented in the form of a power series expansion is quite realistic. The restored expansion coefficients may have a definite physical meaning.

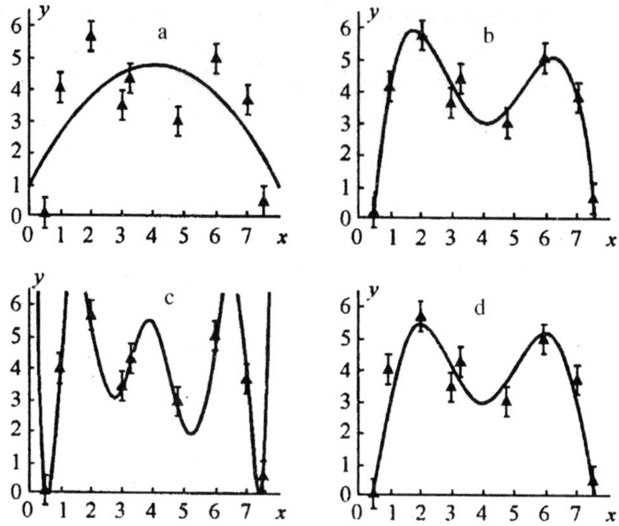

Fig. A2.4 Description of the results of hypothetical experiments by polynomials of the second (*a*), fourth (*b*) and eighth (*c*) degree, curve (*d*) hand drawn.

degree, where the n values of the coefficients were obtained by the method of least squares, passes through all experimental points (Fig. A2.4c). However, this curve does not reproduce the 'character of the dependence' and does not correspond to the error indicated in the graph. The terms, accepted intuitively, are given in quotation marks. There is a well-known theorem according to which for any n points (x_i, y_i), where all x_i differ, there is always a polynomial of the degree $m \leq (n-1)$, and it is the only one passing through all n points. The method of least squares gives exactly that best, in terms of the method of least squares criterion, polynomial passing through all points.

On the basis of the qualitative analysis it may be concluded that in the description, by the theoretical relationships, of the empirical data, this description should be neither 'excessively poor' nor 'excessively good', but be 'in accordance' with the error of the initial data (stricter formulation will be presented in A2.5).

However, the classical methods of restoration of the relationships require either the absolutely accurate description or reaching the minimum of the target function, for example, the minimum of the sum of the squares (or the sum of the modules) of deviations.

A.2.5. DISTRIBUTION OF χ^2

Let $u_i^* = u^*(x_i)$ be the random quantity, *i.e.* measuring $u_i^* = u^*(x_i)$ many times at the given x_i, we obtain different values because of the presence of random errors $\rho_i \equiv \delta u_i = u_i^* - u_i$ (which are also random quantities). Usually, the random error ρ (the index i is omitted) is the sum of a large number of random quantities. Consequently, because of the central limiting Laplace theorem (see [450]), this error is distributed in accordance with the normal (Gauss) law:

$$f(\rho) = \frac{1}{\sqrt{2\pi}\sigma} \exp\left(-\frac{\rho^2}{2\sigma^2}\right), \tag{A2.18}$$

where σ is the mean quadratic or standard deviation; σ^2 is the quantity to which the mean value ρ^2 tends when the number of tests tends to infinity (at some given value of i). The mean value ρ in this case tends to zero and, consequently, its mathematical expectation $M(\rho) = 0$, and the dispersion $D(\rho) = \sigma^2$.

In statistics, it is proved [65, 442] that if the independent random quantities $\rho_i = u_i^* - u_i$ ($i = 1,2, \ldots,n$) are distributed at each value of i in accordance with the normal law, the random quantity $S = \rho^2(Az, u)$ (A2.15), representing the sum of the squares n of these quantities, will be distributed in accordance with the law χ^2 'chi square'. The mathematical expectation χ^2, or $M(S)$, is equal to $n-m$, where n is the number of the results of observations (experimental points), and m is the number of parameters defining the solution, including those taking into account our a priori knowledge regarding the type of solution, which are brought from the outside (those may be the parameters of the theoretical model or the parameters, giving a priori some properties of the solution, for example, the degree of smoothness of the solution, etc). These parameters are determined on the basis of fitting of the experimental data u_i^* by the theoretical dependence. It has also been proved that with the probability of 0.95 (*i.e.* in 95% of cases), random quantity S should fit in interval $\left((n-m)-2\sqrt{2(n-m)},(n-m)+2\sqrt{2(n-m)}\right)$.

Thus, the statistics provide an objective answer (in contrast to subjective conclusions of 'good' or 'bad') to the question of the quality of the description of the experimental dependence $u^*(x_i)$. The description is 'not too bad' and 'not very good' if the quantity

$S = \rho^2 (Az, u)$ fits in the just given confidence interval. Ideally, it should be attempted to bring S as close as possible to $n - m$ (but not to zero!).

A.2.6. CONCEPT OF REGULARISATION

The conditions of adequate description of the results of observations can be fulfilled on the basis of application of the methods of regularisation of solutions of equations of type (A2) proposed and developed in detail by A.N. Tikhonov, V.K. Ivanov, M.M. Lavrent'ev and their successors. On the one hand, these methods take into account adequately the statistical nature of the resultant information and, on the other hand, using the minimum of a priori information, in contrast to modelling methods $z(t)$ when, using a large volume of a priori information on the form of $z(t)$ it is possible to write the hypothetical function $z'(t, a_1, a_2, ..., a_m)$, in a general case non-linear in relation to a_k, modelling $z(t)$. Unknown are only the values of parameters a_k determined on the basis of the least squares criterion using the mathematical method of solving non-linear parametric problems (see Appendix 1).

It should be mentioned that because of a shortage of a priori information, some characteristic features of $z(t)$ (however very important) may already be omitted at the stage of constructing the model. The methods of regularisation of the solution do not require any preliminary information on the form of $z(t)$. It is only necessary to know the response operator A. For linear ill-posed problems, the only assumption is the assumption on the linear nature of mixing of information, *i.e.* on the strict linearity of operator A.

What is the concept of regularisation of the solution? We have already mentioned that product Az should not be excessively sensitive to random changes of the results of observations. In order to fulfil this condition, the authors of the concept of regularisation proposed to impose the smoothness condition on the not-yet-determined solution of $z(t)$. A special smoothing functional $M(A, z, u)$ is introduced. Condition $\rho^2(Az, u) = \min$ is replaced by the condition:

$$M(A,z,u) = \rho^2(Az,u) + \alpha\Omega(z) = \min, \qquad \text{(A2.19)}$$

where $\Omega(z)$ is the stabilising or smoothing functional, α is the parameter of regularisation or the parameter of smoothness of the solution which ensures matching of the type of solution with the error of the initial data.

One of the simplest stabilising functionals is the functional of type

$$\Omega(z)=\sum_{j=2}^{m-1}\left(z_{j-1}-2z_j+z_{j+1}\right)^2, \qquad \text{(A2.20)}$$

which is the sum of squares of the second derivative in finite differences, calculated with the given step. The operator actually characterises the smoothness (non-bending) of the curve. If z is the linear function and the section $[a, b]$ is divided into equal intervals, then evidently $z_{j-1} + z_{j+1} = 2z_j$ and $\Omega(z) = 0$. As the curve becomes less smooth, the value of the operator increases.

At $\alpha = 0$ equation (A2.19) coincides with the requirement of the minimum of the sum of the squares of deviations of theory and experiment. As the value of the regularisation parameter α increases, the 'weight' of the smoothing operator in equation (A2.19) also increases. At a higher value of the parameter α the non-smooth dependences $z(t)$, which increase the generalised discrepancy $M = M(A, z, u)$, are automatically 'cut off'. The best solution is obtained when the generalised discrepancy $M = M(A, z, u)$ approaches $s = n - (m + 1)$ (see paragraph A2.5 of this appendix). The unity is added because in addition to the m values of z_j we also determine the regularisation parameter α.

The procedure for finding the solution is thus reduced to the following. As in the case of determination of the minimum of the relationship (A2.15), it is required that all partial derivatives: $\partial M/\partial z_k = 0$ $(k = 1, 2, \ldots, m)$ are equal to zero. When the problem can be reduced to the matrix form, differentiation gives the system of linear algebraic equations:

$$(B+\alpha C)z = A^*u^*, \qquad \text{(A2.21)}$$

where $B = A^*A$ is the structural or constructional matrix, identical with that formed in the method of least squares, C is a matrix consisting mainly of zeros (only the elements of the main diagonal and its vicinity differ from zero). After finding the solution of system (A2.21), i.e. obtaining z_j, it is necessary to compare $S = \rho^2(Az, u)$ with $s = n - (m + 1)$ and increase or reduce the parameter α. The procedure should be repeated until the value S fits the confidence range in the vicinity of the value s shown in paragraph A2.5 of this appendix (s is the number of degrees of freedom).

Experience with solving different inverse problems shows that when the procedure, matching the accuracy of description of the initial data with their error, is fulfilled, it is possible to describe adequately the experimental data $u^*(x)$ and obtain a 'rational' solution $z(t)$. In the theory of linear ill-posed problems there are proofs of the respective theorems of the convergence of the regularised solution to the true solution, but their analysis is outside the framework of this study.

When operator A is given in the integral form, it is often necessary to use the method of regularisation based on expanding the function $z(t)$ into a series in the range $[a, b]$. This can be carried out by, for example, using the Fourier cosine expansion:

$$z(t) \approx z_N(t) = \sum_{k=0}^{N} a_k \cos\frac{k\pi t}{L}, \tag{A2.22}$$

where $L = a - b$. In this case, the parameter of regularisation is the number of the terms of expansion N. In fact, rejecting high harmonics (when using the criterion of the least squares (A2.15) this is equivalent to adding the smoothing operator, which depends on N, to the sum of squares S), we ensure smoothing of the solution. In this case, S in (A2.15) plays, in fact, the role of M (see equation A2.19). The smoothness of the solution increases with a decrease of N. N is selected using the same scheme as for α. Since the Fourier series is linear in relation to a_k, using the criterion of least squares for finding a_k and solving the resultant system of linear algebraic equations, it is possible to determine the value of $N + 1$ of a_k and, correspondingly, $z_N(t)$. (For this purpose it is necessary to substitute (A2.22) into (A2.12) and require that the partial derivatives: $\partial S/\partial a_k$ ($k = 0, 1, ..., N$) are equal to zero). The value of N, like the value α, is selected by discrepancy S at the number of degrees of freedom $s = n - (N + 1)$. The algorithm of this procedure was described in detail in Appendix 1.

Parameter N may be regarded as continuous because the approximate solution z may be represented by the linear combination of the solutions $z_N(t)$ and $z_{N+1}(t)$ with the 'weights' ensuring the best value of S in the sense of criterion χ^2.

One of the simplest and, at the same time, most effective methods of regularisation has been proposed by M.M. Lavrent'ev.

He showed that the following equation has regularising properties:

$$(A+\alpha E)z=u^*. \tag{A2.23}$$

where E is the unit matrix, *i.e.* matrix whose elements E_{ij} are equal to zero at $i \neq j$ and unity at $i = j$. In other words, to obtain the regularisation effect, it is necessary to add to the diagonal elements of the matrix A the same quantity α whose optimum value is selected as in (A2.21).

In conclusion, we return to Fig. A1.7 which shows the peaks ($-1/2 \rightarrow -3/2$) and ($+1/2 \rightarrow +3/2$) of the Mössbauer spectrum of an iron alloy with 10 at.% Al, with a BCC lattice.

The methods of solving the inverse problems make it possible to transfer from the experimental (instrument) spectrum $\varepsilon(v)$ to the true one $p(x)$ (where $\varepsilon(v)$ has the meaning of $u^*(x)$ and the function $p(x)$ is equivalent to $z(t)$), having a considerably higher resolution in comparison with $\varepsilon(v)$. The relationship (A2.22) is the analogue of (A1.16).

For introductory and extended information on the inverse problems posing and solving, it may be useful to see the monographs [452, 453].

Appendix 3

Relationship of the parameter of pair correlation with the mean effective size of the antiphase domain

In this appendix, we present the results of [63] linked directly with the analysis of the short- and long-range order in crystals using the Mössbauer effect. In this study, for alloys of type *AB*, in which the atoms of one sublattice are surrounded only by the atoms of another sublattice, accurate relationships were obtained between the parameter of pair correlation ε_{AB} (probability characteristic), giving the fraction of the pairs of atoms *AB* in the alloy, and the topological characteristic – the mean effective size of the antiphase domain $<t>$. This confirms the equivalence of the probability and topological characteristic, such as the short-range order and, consequently, the equivalence of the homogeneous and microdomain models of the short-range order.

According to the classification proposed in [61, 454], several morphological types of short-range atomic order are named: homogeneous, microdomain, local and some others.

The initial considerations regarding one or the other type of short-range order (SO) predetermine the specifics of the order parameters used for its description. For example, a homogeneous SO in alloys *AB* is usually characterised by the number of pairs of atoms N_{AA}, N_{AB} and N_{BB}, or by the parameters of pair correlation ε_{AA}, ε_{AB} and ε_{BB}, linked linearly with the number of these pairs. In the case of formation of a SO with a distinctive microdomain structure, the parameter such as the size of the antiphase domain (APD) is used. At that, it is often difficult to find any relationship between different types and the respective parameters of the SO. However, the latter does not mean that there is no such relationship. In fact, different methods of description and the parameters of the SO should be in agreement with each other in limiting cases (for example, when one type of order transfers to another, and also in the cases of maximum ordering, decomposition or complete disordering).

It should be mentioned that these problems are examined only very rarely in the literature. On the whole, in the direction of improvement of the methods of description of SO in the crystals there is obviously a large amount of work to be done, all the more so taking into account the fact that the real structure of the SO is usually more complicated than the models used for this purpose. It should also be mentioned that serious problems form when describing the mutual transformation of the short-range and long-range atomic order, and also in interpretation of the data of local-nuclear methods of investigation of the distribution of atoms in solids (in particular, the Mössbauer effect), not linked with the visualisation of the spatial structure of the atomic order in the direct or inverse space.

In [61, 455] the authors examined theoretically the problem related to the values of the Cowley-Warren SO parameters α_i (measured by the x-ray method as crystal mean values) for the case in which the crystal is a set of almost completely ordered regions of the same composition with the same sufficiently high degree of the short-range order α_{0i} separated by antiphase boundaries (APB). Approximate equations were derived linking the mean over the crystal values of α_i with α_{0i} and the parameters describing the fraction of the atoms on the APB and in the volume of the APD, i.e. expressing α_i through α_{0i} and some effective size of the APD. This equation described some actually observed effects in the formation of the short-range order in crystals.

At the same time, it would be useful to obtain, at least for some relatively simple cases, accurate equations in which the parameters of pair correlation (or the Cowley-Warren parameters linked with them) are found only in the left part (taking also into account the fact that the values of α_{0i} in individual APDs are not known prior to the experiment), and the right hand part would contain only some mean size of the APD and the mean concentration. It is also desirable that these equations describe both short-range ordering and short-range decomposition and do not impose any restrictions on the possibility of fluctuations of the composition.

In [63] derived relationships linking the parameters of the models of homogeneous and microdomain short-range order for a simple case of the type *AB* alloys in which the nodes of one sublattice are surrounded only by the nodes of another sublattice.

A defect-free alloy of the type *AB* was examined. Since the antiphase boundary cannot break inside a perfect crystal, *i.e.* it is

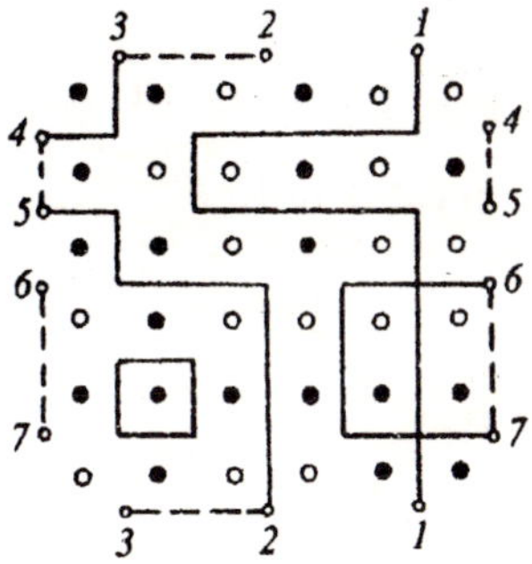

Fig. A3.1. Example of divison of a non-ordered crystal of the *AB* type (two-dimensional case) into fully ordered APD. Population of the sites is determined by flipping a coin. Dash line shows the sections of APB formed during cyclic closure of the lattice

either closed or 'continues to infinity',[1] the alloy with an arbitrary degree of ordering, including the ideally disordered alloy (see Fig. A3.1) may be represented as consisting of completely ordered antiphase domains. In this case, it is assumed that the APD may consist of one or a larger arbitrary number of Casper's polyhedrons. An increase in the degree of order within the scope of such a model indicates an increase of the mean size of the APD tending to infinity in the limit of the total long-range order.

Let us introduce the concept of the effective mean size of the APD as follows:

$$\langle t \rangle = V / S \tag{A3.1}$$

where S is the length of the APB (for a plane lattice it has the dimension of length, for a spatial lattice the dimension of area), V is the volume of the crystal (dimension m^2 or m^3, respectively).

The mean effective size of the domain, determined by this procedure, has a quite concrete geometrical meaning: it is evident that with the accuracy to the shape factor it is the mean minimum size of the domain (diameter for 'spherical' and 'cylindrical' domains, thickness for plane ones, etc).

Let us determine the parameters of paired correlation, describing the numbers of pairs *AA*, *AB* and *BB*, by the conventional procedure:

$$\varepsilon_{AA} = \rho^{AA} - c_A^2; \quad \varepsilon_{AB} = \rho^{AB}/2 - c_A c_B; \quad \varepsilon_{BB} = \rho^{BB} - c_B^2, \tag{A3.2}$$

where $\varepsilon_{AB} = -\varepsilon_{AA} = -\varepsilon_{BB}$ and $\rho^{AA} + \rho^{AB} + \rho^{BB} = 1$,[1] where $\rho^{AA} =$

[1]It should be mentioned that if the finite fragment of the lattice is cyclically closed (a toroid in a flat case), the existing APD are closed and all antiphase domains have finite sizes).

$\frac{N_{AA}}{N}$, $\rho^{AB} = \frac{N_{AB}}{N}$ and $\rho^{AB} = \frac{N_{BB}}{N}$ are the probabilities of corresponding pairs of the closest atoms in the crystal lattice.

Let us express the value of S in terms of the probability of pairs of atoms. Since each pair of the atoms AA or BB of the same name corresponds to the unit section of the APB and vice versa, then

$$S = S_0 \frac{zN}{2}\left(\rho^{AA} + \rho^{BB}\right), \tag{A3.3}$$

where S_0 is the area of the antiphase boundary, belonging to the pair of the atoms AA or BB; N is the total number of lattice sites; z is the coordination number. Taking into account that the volume of the crystal is $V = V_0 N$ (V_0 is the volume per one atom), we obtain:

$$\langle t \rangle = \frac{2}{z}\frac{V_0}{S_0}\frac{1}{\rho^{AA} + \rho^{BB}}. \tag{A3.4}$$

If the measurement unit were represented by $\frac{2}{z}\frac{V_0}{S_0}$, then for a crystal with any lattice taking the above restrictions into account we obtain:

$$\langle t \rangle = \frac{1}{c_A^2 + c_B^2 - 2\varepsilon_{AB}}. \tag{A3.5}$$

Here $c_A c_B \le \varepsilon_{AB} \le c^2_{\max}$ ($c_{\max} = \max\{c_A, c_B\}$).

Let us analyse the resultant relationship. It should be noted that the above examination, taking into account paired correlation effects at the nearest interatomic distances, corresponds to the ordering of type AB. With full ordering, which is possible only at $c_A = c_B$, we have $\varepsilon_{AB} = (1/2) - c_A c_B$ and, consequently, $\langle t \rangle \to \infty$. At complete decomposition $\rho^{AB} \to 0$ and $\varepsilon_{AB} \to -c_A c_B$, we obtain $\langle t \rangle \to 1/(c_A + c_B)^2 = 1$. The last result is fully understandable because in the case of complete decomposition (into regions with 100% content of pure components) the APD will contain, in accordance with its definition, only one atom A or B.

It should be mentioned that when describing the ordering of a more complicated type (for example, A_3B, etc) it is necessary to take into account the effects of atomic correlation at the second

---

[1] $\rho^{AB} = P_{AB} + P_{BA} = 2P_{AB}$ (see (4.11))

and more remote coordination spheres. In this case it is necessary to solve the problem of defining the concept of the APD.

Thus, for the simplest case of ordered alloys of the type *AB* in which the atoms of one sublattice are surrounded by the atoms of another sublattice only, we have shown the equivalence of the models of homogeneous and microdomain SO (and, correspondingly, probability ε_{AB} and topological $<t>$ parameters of the SO). It should be stressed that in this case the pair correlation moments on the first coordination sphere do not give anything else with the exception of $<t>$, *i.e.* they do not provide any information on the form and differences in the dimensions of the domains. The inverse relationships have the form:

$$\varepsilon_{AB} = \frac{1}{2}\left(c_A^2 + c_B^2 - \langle t\rangle^{-1}\right); \quad \rho^{AA} = \frac{1}{2}\left(c_A^2 - c_B^2 + \langle t\rangle^{-1}\right);$$

$$\rho^{AB} = 1 - \langle t\rangle^{-1}; \quad \rho^{BB} = \frac{1}{2}\left(-c_A^2 + c_B^2 + \langle t\rangle^{-1}\right). \qquad \text{(A3.6)}$$

A corresponding relationship may be written for the Cowley–Warren parameters in the first coordination sphere $\alpha_1 (\alpha_1 = -\varepsilon_{AB}/c_A c_B)$.

The relationships (A3.5), (A3.6) simplify the analysis of transition from the liquid SO through an equivalenet microdomain model to a long-range order. In conclusion, it should also be mentioned that the values of ε_{AB} and $<t>$ can be measured by independent experimental methods, for example, ε_{AB} using the Mössbauer effect, and $<t>$ by electron microscopy so that the obtained equations could be verified by experiments.

References

Chapters 1-4

1. Mössbauer R.L., Zeitschrift fur Phys. 1958, **151**, 124-143.
2. Mössbauer R.L., Zeitschrift fur Naturforsch. 1959, **14*a***, 211-216.
3. Mössbauer effect, Collection of articles, translated from English and edited by Yu.M. Kagan, Moscow, IL, 1962.
4. Mössbauer effect, Collection of articles, translated from English and edited by N.A. Burgov and V.V. Sklyarevskii, Moscow, Atomizdat, 1969.
5. Litvinov V.S., Karakishev S.D., Ovchinnikov V.V., Nuclear gamma-resonance spectroscopy of alloys, Moscow, Metallurgiya, 1982.
6. Irkaev S.M., Kuz'min R.N., Opalenko A.A., Nuclear gamma resonance in crystals, Moscow State University, Moscow, 1970.
7. Wertheim G.K., Mössbauer effect, Moscow, Mir, 1966.
8. Belozerskii G.N., Mössbauer spectroscopy as a method of surface study, Energoatomizdat, Moscow, 1990.
9. Kagan Yu., Introductory article, in: Mössbauer effect, Moscow, IL, 1962.
10. Gonser U., Z. Metallkunde, 1966, **57**, 2, 85–96.
11. Maradudin A., Defects and vibrational spectrum of crystals, Moscow, Mir, 1968.
12. Shpinel' V.S., Resonance of x-rays in crystals, Moscow, Nauka, 1969.
13. Gol'danskii V.I., Makarov E.F., Fundamentals of gamma-resonance spectroscopy, in: Chemical applications of Mössbauer spectroscopy, Moscow, Mir, 1970, 9–14.
14. Lugwig G.W. Woodbury H.H., Phys. Rev., 1960, **117**, 5, 1286-1287.
15. Hanna S.S., Heberle J., Littlejohn C., Perlow G.J., Preston R.S. Vincent D.H., Phys. Rev. Letters, 1960, **4**, 4, 177–180.
16. Kvashnina L.B., Krivoglaz M.A., FMM, 1967, **23**, 1, 3-14.
17. Bokshtein B.S., Klinger L.M., Razumovskii I.M., FTT, 1977, **19**, 2, 476-479.
18. Bokshtein B.S., Diffusion in metals, Moscow, Metallurgiya, 1978.
19. Iveronova V.I., Katsnel'son A.A., Izv. VUZov, Fizika, 1976, 8, 40–52.
20. Stearns M.B., J. Magn. and Magn. Mater., 1973, 19th AIP Ann. Conf., Boston, Mass., 1973, 1, N.Y.11.
21. Stearns M.B., Wilson S.S., Measurements of oscillations of the spin density of conduction electrons in ferromagnetic alloys, in: Mössbauer effect, Moscow, Atomizdat, 1969, 247–254
22. Stearns M.B., Phys. Rev. (B), 1972, **6**, 9, 3326-3331.
23. Huffman G.P., Fisher R.M., J. Appl. Phys., 1967, **38**, 2, 735-742.
24. Litvinov V.S., Dorofeev G.A., Ovchinnikov V.V., FMM, 1979, **48**, 2, 426–428.
25. Alekseev L.A., Gruzin P.L., Rodionov Yu.L., Determination of the degree of the short-range order in alloys by nuclear gamma resonance (NGR) method, in: Proc. 4th National conference on ordering of atoms and its effect on the properties of alloys (Tomsk, 1972), Tomsk, 1974, 109-114.
26. Walker L.R., Wertheim G.K., Jaccarino V., Phys. Rev. Lett., 1961, **6**, 98-101.
27. Watson R.E., Freeman A.J., Phys. Rev., 1961, **123**, 6, 2027-2047.

28. Watson R., Distribution of charge and spin density of conduction electrons in metals, in: Hyperfine interactions in solids, translated from English, ed. E.A. Turov, Moscow, Mir, 1970.
29. Stetsenko P.N., Surikov V.V., FTT, 1980, **22**, 7, 1921-1924.
30. Ovchinnikov V.V. Method for determination of individual magnetic moments of atoms using magnetisation and NGR (NMR) data. International conference on Application of the Mössbauer effect (ICAME-1983), Proceedings, Alma-Ata, 1983.
31. Ovchinnikov V.V., Gel'd P.V., FMM, 1988, **65**, 1, 23-31.
32. Stearns M.B., Wilson S.S., Phys. Rev. Lett., 1964, **18**, 9, 313-314.
33. Stearns M.B., J. Appl. Phys., 1965, **36**, 913.
34. Overhauser A.W., Stearns M.B., Phys. Rev. Lett., 1964, **13**, 9, 316-318.
35. Wertheim G.K., Jaccarino V., Wernik J.H., Buchanan D.N.E., Phys. Rev. Lett., 1964, **12**, 1, 24-27.
36. Overhauser A.W., Stearns M.B., Spin susceptibility of conduction electrons in metals with impurities, in: Mössbauer effect (translated from English), Moscow, Atomizdat, 1969, 253-259.
37. Sawer V.E., Reynik R., J Appl. Phys., 1971, **42**, 1604-1617.
38. Campbell I.A., Proc. Phys. Soc., 1966, **89**, 1, 71-76.
39. Vincze I., Campbell I.A., J. Phys. F., 1973, **3**, 3, 647-663.
40. Izyumov Yu.A., Medvedev M.V., Theory of magnetically ordered crystals with impurities, Moscow, Nauka, 1970.
41. Low G.G., Advanced Phys., 1969, **74**, 18, 371-400.
42. Low G.G., Collins M.F., J. Appl. Phys., 1963, **34**, 4, 1195-1199.
43. Collins M.F., Low G.G., J. Phys., 1964, **25**, 5, 596-600.
44. Collins M.F., Low G.G., Proc. Phys. Soc., 1956, **86**, 3, 535-548.
45. Men'shikov A.Z., Spin density of crystals with impurities, Moscow, VINITI, 1974.
46. Stearns M.B., Phys. Rev., B, 1972, **6**, 9, 3326-3331.
47. Kittel Ch., Quantum theory of solids, Moscow, Nauka, 1967.
48. Marshall W., J. Phys (Proc. Phys. Soc.), 1968, Cl, 1, 88-101.
49. Rastogi P.K., Metals and Mater., 1972, **6**, 9, 359-366.
50. Litvinov V.S., Ovchinnikov V.V., Dovgopol S.P., Karakishev S.D., FMM, 1979, **47**, 1, 96-107.
51. Bogachev I.N., Karakishev S.D., Litvinov V S., Ovchinnikov V.V., Izv. VUZov, Fizika, 1974, 6, 83-87.
52. Dorofeev G.A., Litvinov V.S., Metallofizika, 1982, **4**, 3, 38-45.
53. Dorofeev G.A., Ovechkin L.V., Elsukov E.P., Barinov V.A., FMM, 1993, **76**, 4, 107-113.
54. Ryzhenko B.V., Goloborodskii B.Yu., Zaborov A.V., Gel'd P.V., FMM, 1984, **58**, 6, 1153-1163.
55. Ryzhenko B.V., Goloborodskii B.Yu., FTT, 1985, **27**, 2, 421-427.
56. Bashkirov Sh.Sh., Kurbatov G.D., Makhnev E.S., Chistyakov V.A., VAN SSSR, Khimiya, 1975, **223**, 3, 222-624.
57. Sidorenko F.A., FTT, 1981, **23**, 11, 3514-3515.
58. Sidorenko F.A., Gel'd P.V., Many-particle correlations of atomic distribution in the Mössbauer determination of the short-range order parameters in crystals, In: International conference on the application of the Mössbauer effect (ICAME-1983), Proceedings Alma-Ata, Nauka Kaz, SSR, 1983.
59. Ovchinnikov V.V., Zvigintsev N.V., Litvinov V.S., Osminkin V.A., FMM, 1976, **42**, 2, 310-317.

60. Brummer A., Drager G., Mistol J., Annalen der Phys. Folge 7, 1972, **28**, 2, 135-140.
61. Iveronova V.I., Katsnel'son A.A., Short-range order in solid solutions, Moscow, Nauka, 1977.
62. Krivoglaz M.A., Metallofizika, 1984, **6**, 1, 3-36.
63. Ovchinnikov V.V., FMM, 1981, **51**, 5, 828-830.
64. Losievskaja S.A., Phys. Stat. Sol. (a), 1973, **16**, 2, 647-654.
65. Hudson D., Statistics for physicists, translated from English, Moscow, Mir, 1970.
66. Losievskaja S.A., Puzei I.M., Study of Mössbauer spectra of ordered Fe-Al alloys after different heat treatment, In: Proceedings of the 4th National conference on ordering of atoms and its effect on the properties of alloys (Tomsk, 1972), Tomsk, 1974, 169-174.
67. Ovchinnikov V.V., Litvinov V.S., Karakishev S.D., Using the nuclear gamma-resonance method for investigation of ordering in iron alloys, In: Methods of improving the quality of high-strength steels, Urals Polytechnical Institute, Sverdlovsk, 1973, 20-23.
68. Bogachev I.N., Karakishev S.D., Litvinov V.S., Ovchinnikov V.V., Phys. Stat. Sol. (a), 1974, **24**, 2, 661-665.
69. Litvinov V.S., Ovchinnikov V.V., Study of ordering in the Fe-Si system by nuclear gamma-resonance method, In: Physics of metals and metal science, Leningrad Polytechnical Institute, Leningrad, 1975, 42-45.
70. Bogachev I.N., Karakishev S.D., Ovchinnikov V.V., Dorofeev G.A., Study of ordering in transformer steel by nuclear gamma resonance method, In: Proceedings of the 4th National conference on ordering of atoms and its effect on the properties of alloys (Tomsk, 1972), Tomsk, 1974, 201-208.
71. Ovchinnikov V.V., Using the nuclear gamma-resonance method for examining the atomic order in iron alloys, PhD Theses, Urals Polytechnical Institute, Sverdlovsk, 1977.
72. Geichenko V.V., Danilenko V.V., Smirnov A.A., FMM, 1962, 13, 3, 321-332.
73. Inden G., Pitsch W., Chem. Met. Iron and Steel, London, 1973, 314-316.
74. Bykov G.A., Fam Zui Khien, ZhETF, 1962, **43**, 3, 909-912.
75. Abezgauz G.G., Tron' A.P., Kopenkin Yu.N., Korovina I.A., A handbook of probability calculations, Moscow, Voenizdat, 1970.
76. Krivoglaz M.A., Smirnov A.A., Theory of ordering alloys, Moscow, FM, 1958.
77. Sidorenko F.A., Pervukhin N.A. Ryzhenko B.V., FTT, 1983, **25**, 8, 2257-2263.
78. Dormann J.L., Brossard L., Fatseas G.A., Phys. Stat. Sol (b) 1972, **52**, 1, K23-K26.
79. Ovchinnikov V.V., Development and application of the Mössbauer spectroscopy method for analysis of the atomic and magnetic structure of alloys, Dr. Sci. Theses, Urals Polytechnical Institute, Sverdlovsk, 1988.
80. Stearns M., Variation of the spin density and spin susceptibility of 4-*s* conduction electrons in iron, In: Wertheim H., Mössbauer effect, translated from English, Moscow, Mir, 1966, 128-133.
81. Ovchinnikov V.V., Romashev L.N., Calculating the configurations of the local environment of ^{57}Fe atoms and concentration profiles of interfaces in ^{57}Fe/Cr superlattices by the Monte-Carlo method, In: Proceedings of the 16th International School-Seminar "New Magnetic Materials for Microelectronics",

Moscow, Moscow State University, 1998, 35-36.

82. Ovchinnikov V.V., Ustinov V.V., Romashev L.N., Goloborodskii B.Yu., Semerikov A.V., Calculating the configurations of the local environment of ^{57}Fe atoms in $^{57}Fe/Cr$ superlattices by the Monte Carlo method, modelling of the Mössbauer spectra and experimental investigation of the effect of ion irradiation on the giant magnetoresistive effect, In: Proceedings of the 11th International Seminar "Radiation physics of solids" (Sevastopol, June 25-30, 2001), edited by Bondarenko G.G., Moscow, NII PMT MGIEM (TU), 2001, 254-261, ISBN 5-89671-002-9.

Chapter 5

83. Ono K., Ishikawa I., Ito A., J. Phys. Soc. Japan, 1962, **17**, 11, 1747-1750.
84. Cser L., Ostanevich I., Pal L., Phys. Stat. Sol., 1967, **20**, 2, 581-589.
85. Lessoille M.R., Gielen P.M., Phys. Stat. Sol., 1970, **37**, 1, 127-139.
86. Kuz'min R.N., Losievskaja S.A., FMM, 1970, **29**, 3, 569-577.
87. Glezer A.M., Molotilov B.V., Izv. AN SSSR, Metally, 1972, 4, 172-177.
88. Shinjo T., Nakamura Y., J. Phys. Soc. Japan, 1963, **18**, 6, 797-801.
89. Stearns M.B., Phys. Rev., 1963, **129**, 3, 1136-1144.
90. Gustin C., Gielen P.M., Croylaan G., Scr. Met., 1970, **74**, 11, 925-928.
91. Papadimitriou G., Genin J.M., Phys. Stat. Sol. (a) 1972, **9**, 1, K19-K23.
92. Dorofeev G.A., Ovchinnikov V.V., Segal' V.M., Shestakova G.A., FMM, 1979, **48**, 3, 669-672.
93. Sachkov I.N., Chemerinskaya L.S., Ovchinnikov V.V., FTT, 1996, **38**, 2, 603-608.
94. Bandyopadhyay D., Suwas S., Singru R.M., Bhargava S., J. Mater. Sci., 1998, **33**, 1, 109-116.
95. Miyazaki Masaburo, Ichikawa Masashi, Komatsu Takayuki, Matušita Kazumasa, Nakajima Kensuke., J Appl. Phys., 1991, **69**, 3, 1556-1563.
96. Mohan T.V.S.M., Bansal Babu C., Phys. Stat. Sol. B, 1996, **193**, 1, 167-175.
97. Satula D., Dobrzynski L., Waliszewski J., Szymanski K., Reko K., Malinowski A., Bruchel Th., Sharpf O., Blinowski K., J. Magn. and Magn. Mater., 1997, **169**, 3, 240-252.
98. Moriya Takeshi, Nakashima Hiroyuki, Isokane Yuichiro, Miyazaki Totu, Kozakai Takao, Koyama Toshiyuki., J. Phys. Soc. Jap., 1996, **65**, 1, 293-296.
99. Voronin V.I., Serikov V.V., Kleinerman N.M., Kuchin A.G., Physica B, 2000, **276-278**, 570-571.
100. Pradell T., Clavaguera N., Zhu J., Clavaguera-Mora M.T., J. Phys. Condens. Matter, 1995, **7**, 21, 4129-4173.
101. Shabashov V.A., Sagaradze V.V., Morozov S.V., Volkov G.A., Metallofizika, 1990, **12**, 4, 107-114.
102. Sagaradze V.V., Shabashov V.A., Lapina T.M., Pecherkina I.L., Pilyugin V.P., FMM, 1994, **78**, 6, 58-59.
103. Shabashov V.A., Sagaradze V.V., Morozov S.V., Volkov G.A., Lapina T.M., FMM, 1991, 12, 119-129.
104. Elsukov E.P., Structure and magnetic properties of microcrystalline and amorphous alloys of iron with *sp*-elements, Dr. Sci. Theses, Moscow, Moscow State University, 1994.
105. Elsukov E.P., Barinov V.A., Galakhov V.R., Yurchikov E.E., Ermakov A.E., FMM, 1983, **55**, 2, 337-340.

106. Elsukov E.P., Konygin G.N., FMM, 1989, **67**, 2, 301-310.
107. Elsukov E.P., Vorobev Yu. N., Trubachev A.V., Barinov V.A., Phys. Stat. Sol. (a), 1990, **117**, 1, 291-298.
108. Elsukov E.P., Vorobev Yu. N., Trubachev A.V., Phys. Stat. Sol. (a), 1991, **127**, 1, 215-222.
109. Elsukov E.P., Konygin G.N., Barinov V.A., Voronina T.V., J. Phys.: Condens. Matter, 1992, **4**, 7597-7606.
110. Elsukov E.P., Voronina T.V., Barinov V.A., J. Magn. and Magn. Mater., 1992, **115**, 271-280.
111. Elsukov E.P., FMM, 1994, **76**, 5, 5-31.
112. Eluskov E.P., Yakovlev V.V., Barinov V.A., FMM, 1994, **77**, 4, 131-137.
113. Yelsukov E.P., Vorobyov Yu.N., Arbusova T.I., Smolyak I.B., J. Magn. and Magn. Mater., 1994, **130**, 44-50.
114. Voronina E.V., Fomin V.M., Deev A.N., Yelsukov E.P., J. Phys. IV France, 1997, **7**, 2, 1003-1004.
115. Fomin V.M., Voronina E.V., Yelsukov E.P., Deev A.N., Mater. Sci. Forum, 1998, **269-272**, 437-442.
116. Elsukov E.P., Voronina E.V., Fomin V.M., Konygin G.N., FMM, 1998, **85**, 3, 89-99.
117. Voronina E.V., Formin V.M., Babanov Yu.A., Elsukov E.P., Konygin G.N., Godovikov S.K., Izvestiya RAN, Seriya Fizicheskaya, 1999, **63**, 7, 1430-1434.
118. Fomin V.M., Voronina E.V., Elsukov E.P., Konygin G.N., Vestn. Udm. un-ta, 1997, 4, 58-67.
119. Marchal G., Mangin Ph., Piecuch M., Janot Chr., J. Phys., 1976, **37**, 12, 763-768.
120. Bansal C., Campbell S.J., Stewart A.M., J. Magn. and Magn. Mater., 1982, **27**, 195-201.
121. Rusakov V.S., Mössbauer spectroscopy of locally inhomogeneous systems, Dr. Sci. Theses, Moscow, Moscow State University, 1999.
122. Rusakov V.S., Kadyrzhanov K.K., Turkebaev T.E., Aimanov M.Sh., Zhukov V.N., Poverknost'. Fizika, Khimiya, Mekhanika, 1995, 7-8, 28-35.
123. Rusakov V.S., Kadyrzhanov K.K., Turkebaev T.E., Aimanov M.Sh., Zhukov V.N., Poverkhnost': Fizika, Khimiya, Mekhanika, 1996, 11, 80-90.
124. Yagova O.E., Rusakov V.S., Kuzmina N.A., Bychkov A.M., Experiment in Geosciences, 1997, **6**, 2, 45-46.
125. Kupin Yu. G., Rusakov V.S., Badyukov D.D., Kozlov E.A., Experiment in Geosciences, 1997, **6**, 2, 62-63.
126. Kuzmina N.A., Bychkov A.M., Rusakov V.S., Chistyakova N.I., Experiment in Geosciences, 1997, **6**, 2, 92-93.
127. Kotelnikova A.A., Rusakov V.S., Bychkov A.M., Experiment in Geosciences, 1997, **6**, 2, 94-95.
128. Kupin Yu. G., Badyukov D.D., Rusakov V.S., Experiment in Geosciences, 1998, **7**, 2, 57-58.
129. Rusakov V.S., Ilyushin A.S., Vinogradova A.S., Nikanorova I.A., Cherepanov V.M., Metally. RAN, 1999, 1, 91-97.
130. Rusakov V.S., Ilyushin A.S., Vinogradova A.S., Nikanorova I.A., Cherepanov V.M., Izvestiya RAN. Seriya Fizicheskaya, 1999, **63**, 7, 1404-1409.
131. Rusakov V.S., Ilyushin A.S., Vinogradova A.S., Nikanorova I.A., Cherepanov V.M., Vestn. Mosk. Un-ta. Ser. 3, Fizika. Astronomiya, 1999, 2, 46-49.
132. Rusakov V.S., Ilyushin A.S., Baranov A.B., Vestn. Mosk. Un-ta. Ser. 3, Fizika.

Astronomiya, 1999, 3, 41-51.
133. Chamberad A., Laugiar J., Penisson J.M., J. Magn. and Magn. Mater., 1979, **10**, 2-3, 129-144.
134. Rodionov Yu.L., Isfandiyarova G.G., Sarsenbin O.S., FMM, 1979, **48**, 5, 979-985.
135. Watanabe D., Moreta H., Saito H., Ogawa S., J. Phys. Soc. Japan, 1970, **29**, 3, 722-729.
136. Dorofeev G.A., Litvinov V.S., FMM, 1977, **44**, 2, 438-440.
137. Gol'dshtein Ya.E., Charushnikova G.A., Izv. AN SSSR. Metally, 1966, 5, 75-82.
138. Ovchinnikov V.V., Litvinov V.S., Charushnikova G.A., FMM, 1979, **47**, 5, 1099-1102.
139. Amigud G.G., Ovchinnikov V.V., Litvinov V.S., Filippov M.A., FMM, 1981, **51**, 5, 955-961.
140. Sidorova G.V., Korneev V.P., Milyaev I.M., et al., Metally, 1997, 6, 90-92.
141. Makhneva T.M., Makhnev E.S., Elsukov E.P., et al., Stabilisation of austenite in high-strength chromium-nickel steel in ageing. Micromechanisms of plasticity, fracture and accompanying phenomena, Proceedings of the International Conference, Tambov, June 24-28, 1996, 191.
142. Cieslak J., Dubiel S.M., J. Alloys and Compounds, 1998, **269**, 1-2, 208-218.
143. Crespo P., Menendez N., Tornero J.D., et al., Acta Mater., 1998, **46**, 12, 4161-4166.
144. Plazaola F., Orue I., Fernandez-Gubieda M.L., Barandiaran J.M., J. Appl. Phys., 1995, **77**, 7, 3338-3342.
145. Gupta A., Kane S.N., Kraus L., Duhaj P., J. Magn. and Magn. Mater., 1995, **140-144**, 1, 321-322.
146. Kalimulin R.K., Korsunskii I.L., Lomonosov V.V., Meryshev A.A., Pazhin Yu.F., Samarin P.F., Silyakov R.S., Skvortsov V.V., Investigation of phase transformations in quenched $Ti_{1-x}Fe_x$ alloys by nuclear gamma resonance method, Moscow, Preprint of IAE-5560/1, 1992.
147. Korablev V.A., Kabes A.I., Nakorneeva T.D., Distribution of chromium atoms in solid solution based on a-iron after g®a martensitic transformation. In: Termicheskaya Obrabotka i Fizika Metallov: Mezhvuz, sbornik. Sverdlovsk, Urals Polytechnical Institute, 1987, 36-40.
148. Kabes A.I., Study of the effect of austenitising on the structure in order to increase the mechanical properties of chromium steels, PhD Theses, Urals State Technical University, Ekaterinburg, 1993.
149. Korablev V.A., Ivchenko V.A., Kabes A.I., et al., FMM, 1990, **63**, 1, 163-170.
150. Zhantikin T.M., Mössbauer spectrosopy of radiation damage of some intermetallics and amorphous iron alloys, PhD Theses, Academy of Sciences of Kazakh SSR, Institute of Nuclear Physics, Alma-Ata, 1987.
151. Ibragimov Sh.Sh., Melikhov V.D., Skakov M.K., Zhetbaev A.K., Vereshchak M.F., Zhantikin T.M., Radiation Effects, 1984, **80**, 1-2, 15-21.
152. Zhantikin T.M., Zhetbaev A.K., Vereshchak M.F., Hyperfine Interactions, 1985, **29**, 1297-1301.
153. Sagaradze V.V., Shabashov V.A., Lapina T.M., Arbuzov V.L., FMM, 1994, **78**, 4, 88-96.
154. Sagaradze V.V., Lapina T.M., Shabashov V.A., Arbuzov V.L., FMM, 1997,

83, 5, 121-126.

155. Bugaev V.N., Gavrilyuk V.G., Nadutov V.M., Tatarenko V.A., FMM, 1989, **68**, 5, 931-940.
156. Gavrilyuk V.G., Nadutov V.M., Gladun O.V., FMM, 1990, **9**, 3, 128-134.
157. Nadutov V.M., Tatarenko V.A., Tsynman C.L. Ullakko K., Journal of Metal Physics and Advanced Technologies, 1994, **16**, 8, 34-40.
158. Berns H., Duz V.A., Ehrhardt R., et al., Z. Metallk., 1997, **88**, 2, 109-116.
159. Gavriljuk V., Nadutov V., Cryogenics, 1994, **34**, ICEC Suppl., 485-488.
160. Shabashov V.A., Mamaev S.V., Volkov G.A., FMM, 1993, **75**, 1, 54-58.
161. Bugaev V.M., Gavrilyuk V.G., Gladun O.V., Efimenko S.P., Nadutov V.M., Tatarenko V.A., Distribution of carbon and nitrogen atoms in autstenite, In: High-nitrogen steels, Proceedings of the 1st National Conference, Kiev, April 18-20, 1990, Kiev, 147-162.
162. Karakishev S.D., FMM, 1995, 3, 95-100.
163. Schastlivtsev V.M., Tabatchikova T.I., Yakovleva I.L., et al., FMM, 1996, **82**, 6, 102-115.
164. Fall I., Genin J.-M.R., Met. and Mater. Trans. A, 1996, **27**, 8, 2160-2177.
165. Ullakko K., Gavriljuk V.G., Nadutov V.M., Met. and Mater. Trans. A, 1994, **25**, 5, 889-909.
166. Ovchinnikov V.V., Litvinov V.S., Zvigintsev N.V., Izvestiya VUZov. Fizika, 1981, 12, 29-33.
167. Tereshchenko N.A., Shabashov V.A., Uvarov A.I., FMM, 1997, **83**, 5, 96-103.
168. Shabashov V.A., Nikolaev A.L., Mukoseev A.G., Sagaradze V.V., Filippova N.P., Izvestiya Akademii Nauk, Seriya Fizicheskaya, 2001, **65**, 7, 1010-1015.
169. Dubovtsev I.A., Grazhdankin V.N., Vereshkov G.M., Losev N.F., Izv. Severo-Kavkazskogo Nauchnogo Tsentra Vysshei Shkoly, 1987, 3, 84-87.
170. Dubovtsev I.A., Grazhdankin V.N., Vereshkov G.M., Losev N.F., Izvestiya Akademii Nauk, Seriya Fizicheskaya, 1988, **52**, 9, 1711-1715.
171. Mostafa M.A., Balogh Judit, Kuzmann E., Phys. Stat. Sol., 1986, **A96**, 2, 445-449.
172. Guimaraes R.B., Viccaro P.J., Schreiner W.H., Vasconcellos M.A.Z., Baibich M.N., J. Phys., 1988, **49**, 12, Suppl. 2, 1375-1376.
173. Kuzmann E., Vertes A., Ianov L.A., Novochaitski I.A., Hyperfine Interactions, 1989, **45**, 1-4, 309-313.
174. Skorvanek Ivan, Miglierini Marcel., J. Magn. and Magn. Mater., 1991, **96**, 1-3, 125-136.
175. Zych Wlodzimierz, Kopcewicz Michal, Mater. Sci. and Eng., 1991, **134**, 880-882.
176. Orue I., Fdez-Gubicda M.L., Plazaola F., Barandiaran J.M., J Phys. Condens. Matter. (J. Phys. F), 1998, **10**, 17, 3807-3822.
177. Dmitrieva N.V., Kleinerman N.M., Lukshina V.A., Serikov V.V., Potapov A.P., Thermal stability of the induced magnetic anisotropy and structure of the nanocrystalline alloy FeCuNbSiB. In: 14th International Symposium on Soft Magnetic Materials (SMM 14). Balatonfured, 8-10 Sept., 1999. J. Magn. and Magn. Mater., 2000, **215-216**, 453-454.
178. Smith P.A.I., Coey J.M.D., J. Magn. and Magn. Mater. 1999, **196-197**, 199-200.
179. Kleinerman N.M., Serikov V.V., Lukshin V.A., Dmitrieva N.V., Potapov A.P., FMM, 2001, **91**, 6, 46-50.

180. Litvinov V.S., Karakishev S.D., Tsurin V.A., FMM, 1977, **43**, 2, 352-357.
181. Karakishev S.D., Litvinov V.S., Poptsov M.E., FMM, 1982, **54**, 4, 811-812.
182. Litvinov V.S., Poptsov M.E., Chumakova L.D., FMM, 1984, **58**, 5, 1037-1039.
183. Kaigorodov V.N., Klotsman S.M., Koloskov V.M., Tatarinova G.N., FMM, 1986, **66**, 5, 958-965.
184. Kaigorodov V.N., Klotsman S.M., Koloskov V.M., FMM, 1987, **63**, 3. 614-616.
185. Kaigorodov V.N., Klotsman S.M., Koloskov V.M., Shlyapnikov S.N., Poverkhnost': Fiz. Khimiya. Mekh., 1987, 3, 124-129.
186. Kaigorodov V.N., Klotsman S.M., Koloskov V.M., Shlyapnikov S.N., FMM, 1988, **66**, 1, 200-202.
187. Klotsman S.M., Kaigorodov V.N., FMM, 1999, **87**, 6, 92-100.
188. Klotsman S.M., Kaigorodov V.N., FMM, 1999, **88**, 4, 52-62.
189. Klotsman S.M., Kaigorodov V.N., Kurkin M.I., Dyakin V.V., FMM, 2000, **89**, 5, 93-100.
190. Kaigorodov V.N., Klotsman S.M., FMM, 1999, **87**, 5, 68-77.
191. Rixecker G., Birringer R., Gleiter H., Gonser U., Mössbauer Effect of ^{57}Fe at Grain in Ball Milled Tungsten. Proc. ICAME 95 – paper O4 – C4.
192. Herr U., Jing J., Birringer R., Gonser U., Gleiter H., Appl. Phys. Lett., 1987, **50**, 8, 472-474.
193. Valiev R.Z., Mulyukov R.R. and Ovchinnikov V.V., Phil. Mag. Letters, 1990, **62**, 4, 253-256.
194. Valiev R.Z., Mulyukov R.R., Ovchinnikov V.V., Shabashov V.A., Arkhipenko A.Yu., Safarov I.M., Metallofizika, 1990, **12**, 5, 124-126.
195. Valiev R.Z., Mulyukov R.R., Ovchinnikov V.V., Shabashov V.A., Scr. Met., 1991, **25**, 12, 2717-2722.
196. Ovchinnikov V.V., Chernoborodov V.N., Mikhailischeva E.P., Valiev R.Z., Mulyukov R.R., Amirkhanov N.M., Trans. Mat. Res. Soc., 1994, **16B**, 1489-1492.
197. Shabashov V.A., Ovchinnikov V.V., Mulyukov R.R., Valiev R.Z., Filippova N.P., Phys. of Metals and Metallography, 1998, **85**, 3, 318-326.
198. Oshtrakh M.I., Kopelyan E.A., Semionkin V.A., Livshits A.B., Krylova V.E., Kozlov A.A., Nucl. Instr. Meth. in Phys. Res., 1993, **B76**, 405-407.
199. Oshtrakh M.I., Semionkin V.A., Milder O.B., Nucl. Instrum. Meth. in Phys. Res., 1999, **B151**, 227-231.
200. Oshtrakh M.I., Milder O.B., Semionkin V.A., Berkovsky A.L., Azhigirova M.A., Vyazova E.P., Z. Naturforsch., 2000, **55a**, 193-198.
201. Wertheim G.K., Jaccarino V., Wernik J.H., Buchanan D.N.E., The length of exchange interaction in iron-based alloys, In: Mössbauer effect, translated from English, edited by Yu. Kagan, Moscow, Atomizdat, 1969, 260-269.
202. Karakishev S.D., Litvinov V.S., Amigud G.G., et al., Trudy VUZov RF, Urals Polytechnical Institute, Sverdlovsk, 1973, 94-99.
203. Ovchinnikov V.V., Pervukhin N.A., Ryzhenko B.V., et al., FTT, 1981, **23**, 2, 617-620.
204. Ovchinnikov V.V., Pervukhin N.A., Ryzhenko B.V., et al., FTT, 1981, **23**, 10, 3045-3048.
205. Stearns M.B., Appl. Phys., 1964, **35**, 1095-1096.
206. Khachaturyan A.T., Theory of phase transformations and structure of solid solutions, Moscow, Nauka, 1974.

207. Semenovskaya S.V., Phys. Stat. Sol. B, 1974, **64**, 1, 291-303.
208. Wilde D.J., Methods for extremum finding (transl. from English), Moscow, Nauka, 1967.
209. Ovchinnikov V.V., Development and application of the Mössbauer methods of analysis of the atomic and magnetic structure of alloys, Dr. Sci. Dissertation, Sverdlovsk, 1988.
210. Demidenko V.S., Egorov S.N., FMM, 1993, **75**, 5, 87-93.
211. Delyagin N.N., Gurevich G.M., Erzinkyan A.L., et al., ZhETF, 1996, **109**, 4, 1451-1464.
212. Elsukov E.P., Voronina E.V., Barinov V.A., Konygin G.N., Godovikov S.K., Yakovlev V.V., Zagainov A.V., FMM, 1996, **82**, 4, 64-70.
213. Daniyarov Sh.M., Ovchinnikov V.V., Ryzhenko B.V., Sidorenko F.A., Yuldashev U.Yu., FMM, 1985, **60**, 4, 736-741.
214. Volkova N.V., Kleinerman N.M., Kourov N.I., Naish V.E., Sagaradze I.V., Serikov V.V., Tyulenev L.N., Physics of Metals and Metallography, 2000, **89**, 1, 35-41.
215. Macri P.P., Rose P., Frattini R., Enzo S., Principi G., Hu W.X., Cowlam N. J., Appl. Phys., 1994, **76**, 7, 4061-4067.
216. Jiang J.Z., Chen F.T., J. Phys.: Condens. Matter, 1994, **6**, 26, L343-L348.
217. Ding J., Eilon M., Street R., et al., J. Magn. and Magn. Mater., 1995, **140-144**, 1, 471-472.
218. Jartych E., Zurawicz J.K., Oleszak D., Pekala M., J. Phys.: Condens. Matter. (J. Phys. F.), 1998, **10**, 22, 4929-4936.
219. Larica C., Alves K.M.B., Baggio-Saitovitch E., Guimaraes A.P., J. Magn. and Magn. Mater., 1995, **145**, 3, 306-312.
220. Balddokhin Yu.V., Kolotyrkin P.Ya., Petrov Yu.I., Shafranovskii E.A., Dokl. AN RAN, 1996, **347**, 1, 51-54.
221. Tsurin V.A., Ermakov A.E., Lebedev Yu.G., Filippov B.N., Phys. Stat. Sol. (a), 1976, **33**, 325-332.
222. Tsurin V.A., Fillipova N.P., Men'shikov A.Z., Stepanov A.P., FMM, 1983, **56**, 5, 1036-1037.
223. Tsurin V.A., Sorkin A.M., Barinov V.A., FMM, 1989, **68**, 5, 889-893.
224. Men'shikov A.Z., Tsurin V.A., FMM, 1979, **47**, 2, 305-309.
225. Tsurin V.A., Fillipova N.P., Stepanov A.P., FMM, 1983, **25**, 10, 3072-3077.
226. Tsurin V.A., Fillipova N.P., Stepanov A.P., FMM, 1986, **61**, 6, 1114-1116.
227. Tsurin V.A., Fillipova N.P., Stepanov A.P., FMM, 1987, **29**, 9, 2694-2698.
228. Tsurin V.A., Fillipova N.P., Stepanov A.P., FMM, 1986, **28**, 6, 1813-1817.
229. Li Fa-Shen, Sun Ji-Jun, Chien C.L., J. Phys.: Condens. Matter, 1995, **7**, 9, 1921-1931.
230. Shabashov V.A., Tereshchenko N.A., Uvarov A.I., Lapina T.M., FMM, 1997, **83**, 4, 95-102.
231. Rusakov V.S., Ilyushin A.S., Morozov V.N., Nikanorova I.A., Izv. AN. Ser. Fiz., 1994, **58**, 4, 24-29.
232. Ilyushin A.S., Rusakov V.S., Nikanorova I.A., Avdeeva I.A., Korchazhkin I.V., Graudin'sh E.Ya., FTT (St. Petersburg), 1996, **38**, 9, 2875-2886.
233. Itoh K., Ohkubo N., Kanematsu K., Jap. J. Appl. Phys., Pt. 1, 1993, **32**, Suppl. 32-3, 322-324.
234. Marasinghe G.K., Mishra S., Pringle O.A., Long Gary J., Hu Z., Yelon W.B., Grandjean F., Middleton D.P., Buschow K.H.J., J. Appl. Phys., 1994, **76**, 10, Pt 2, 6731-6733.
235. Long G.J., Marasinghe G.K., Mishra S.R., Pringle O.A., Hu Z., Yelon W.B.,

Middleton D.P., Buschow K.H.J., Grandjean F.A., J. Appl. Phys., 1994, **76**, 9, 5383-5393.

236. Middleton D.P., Mishra S.R., Long G.J., Pringle O.A., Hu Z., Yelon W.B., Grandjean F., Buschow K.H.J., J. Appl. Phys., 1995. **78**, 9, 5568-5576.

237. Sanchez J.P., Blaise A., Tomala K., Kmiec R., Solid State Commun., 1994, **89**, 7, 647-649.

238. Onodera Hideya, Hisao Kobayashi, Hiroshi Yamauchi, Masayoshi Ohashi, Yasuo Yamaguchi, J. Magn. and Magn. Mater., 1997, **170**, 1-2, 201-210.

239. Bonville P., Hodges J.A., Imbert P., Jehanno G., Thuery P., J. Magn. and Magn. Mater., 1994, **136**, 3, 238-244.

240. Onodera Hideya, Nurata Akifumi, Koizuka Masaaki, Ohashi Masayoshi, Yamaguchi Yasuo, Sci. Repts. Res. Inst. Tohoku Univ. A, 1994, **40**, 1, 177-184.

241. Ermakov A.E., Yurchikov E.E., Elsukov E.P., Barinov V.A., Chukalkin Yu.G., FMM, 1982, **24**, 7, 1947-1951.

242. Barinov V.A., Tsurin V.A., Gaviko U.S., Yermakov A.E., Teytel Ye.I., Shegoleva, Leccabue F., Watts B.E., Panizzieren R., Bocelli G., Castanon S., J. Magn. and Magn. Mater., 1995, **139**, 143-150.

243. Maikov V.V., Yermakov A.Ye., Yelsukov Ye.P., Konygin G.N., Leccabue F., Watts B.E., Vazquez M., S. Diaz Castanon S., J. Magn. and Magn. Mater., 1995, **151**, 167-172.

244. Barinov V.A., Dorofeev G.A., Ovechkin L.V., Elsukov E.P., Ermakov A.E., Phys. Stat. Sol. a, 1991, **123**, 527-534.

245. Ermakov A.E., Yurchikov E.E., Barinov V.A., FMM, 1981, **52**, 6, 1184-1193.

246. Bremers H., Fricke M., Hesse Y., Hyperfine Interactions, 1994, **94**, 1855-1859.

247. Bremers H., Jarms Ch., Hesse J., Mössbauer spectra of $Fe_{89-x}Mn_{11}Al_x$ and their description in a local model. Proceeding of ICAME 95, 1995, 1-4.

248. Bremers H., Jarms Ch., Hesse J. Chadjivasiliou S., Efthimiadis K.G., Tsoukalas I. J., J. Magn. and Magn. Mater., 1995, **140-144**, 63-64.

249. Yermakov A.Ye., Uimin M.A., Shangurov A.V., et al., Mat. Sci. Forum, 1999, **225-227**, part 2, 147-156.

250. Wanderka N., Czubayko U., Naundorf V., Ivchenko V.A., Yermakov A.Ye., Uimin M.A., Wollenberg H., Ultramicroscopy, 2001, **89**, 189-194.

251. Ivchenko V.A., Wanderka N., Czubayko U., Naundorf V., Ermakov A.Ye., Uimin M.A. and Wollenberg H., Mater. Sci. Forum, 2000, **343/346**, 709-714.

252. Mushnikov N.V., Goto Tsuneaki, Zajkov N.K., Gaviko V.S., Serikov V.V., Kleinerman N.M., Yermakov A.Ye., J. of Alloys and Compounds, 1999, **284**, 70-76.

253. Mushnikov N.V., Zujkov N.K., Serikov V.S., Kleinerman N.M., Gaviko V.S., Yermakov A.Ye., J. Magn. and Magn. Mater., 1997, **167**, 93-98.

254. Yermakov A.Ye., Zujkov N.K., Mushnikov N.V., Gaviko V.S., Serikov V.S., Kleinerman N.M., Nanostructured Materials, 1999, **12**, 797-780.

255. Sidorova G.V., Novakova A.A., Braude I.S., Fizika i Khimiya Obrabotki Materialov, 1986, 2, 70-72.

256. Novakova A.A., Kiselev A.A., Kuz'min R.N., Sidorova G.V., Pis'ma v ZhETF, 1986, **43**, 7, 321-324.

257. Novakova A.A., Kiselev A.A., Sidorova G.V., Sirotina G.A., Izvestiya AN SSSR. Seriya Fizicheskaya, 1986, **50**, 12, 2449-2454.

258. Buchkov D., Nikolov S., Dragieva I., Slavcheva M., J. Magn. and Magn. Mater., 1986, **62**, 1, 87-92.
259. Morup S., Wonferghem J. van, Meagher A., Koch C.J.W., IEEE Trans Magn., 1987, **23**, 5, pt 2, 2978-2980.
260. Srivastava B.K., Ericsson T., Haggstrom L., Verma H.R., Andersson Y., Rundqvist S. A., J. Phys. C: Solid State Phys., 1987, **20**, 3, 463-472.
261. Barandiaran J.M., Fdez-Gubieda M.L., Plazaola F., Nielsen O.V., J. Phys., 1988, **49**, 12, suppl. 2, 1367-1368.
262. Donzelli O., Fratucello G., Ronconi F., Allia P., Vinai F., Vera A., J. Phys., 1988, **49**, 12, Suppl. 2, 1319-1320.
263. Gwiazda J., Marianska E., Zych W., Phys. Stat. Sol. A, 1988, **108**, 1, K67-K72.
264. Taniwaki M., Maeda M., Sci. and Eng., 1988, **99**, 47-51.
265. Xu Zuxiong, Ma Ruzhang, Acta Metal. Sin., 1988, **24**, 1, B71-B73.
266. Nowik I., Felner I., Wolfus Y., Yeshurun Y., J. Phys. F., 1988, **18**, 9, L181-L185.
267. Lagu Anuradha, Shringi S.N., Nigam A.K., Chandra Girish, Prasad Shiva, Krishnan R. Hyperfine Interactions, 1989, **51**, 1-4, 1025-1030.
268. Yamasaki Tohru, Ogino Yoshihiyo, J. Jap. Inst. Metals, 1989, **53**, 4, 359-366.
269. Kitano Yohko, Maeda Chizuko, Shimizu Masato, Kobayashi Shinji, Morito Nobuyuki. J. Jap. Inst. Metals, 1989, **53**, 3, 258-265.
270. Zemeik T., Jiraskova Y., Hyperfine Interactions, 1989, **51**, 1-4, 1039-1044.
271. Laggoun A., Teillet J., Lassri H., Krishan R., Papaefthymiou G.C., Solid State Commun., 1989, **71**, 2, 79-83.
272. Ressler L., Borgmeier B., Erdmann K., Rosenberg M., Hyperfine Interactions, 1989, **51**, 1-4, 1031-1038.
273. Ganesan K., Narayanasamy A., Nagarajan T., J. Phys. Condens Matter., 1990, **2**, 18, 4227-4242.
274. Dudek W., Gwiazda J., Marianska E., Oleniacz J., Zych W., J. Magn. and Magn. Mater., 1990, **86**, 2-3, 213-218.
275. Cadogan J.M., Day R.K., Dounlop J.B., Foley C.P., Less-common metals, 1990, **158**, 2, L45-L49.
276. Miglierini M., Sitek J., Acta Phys. Slov., 1990, **40**, 4, 254-256.
277. Gupta Ajay, Habibi S., Mater. Sci. and Eng. A, 1991, **133**, 375-378.
278. Rogalski M.S., Jackson T.J., Bibicu I., Palmer S.B., J Phys. D, 1994, **27**, 10, 2167-2170.
279. Fiorani D., Malizia F., Ronconi F., Testa A.M., J. Magn. and Magn. Mater., 1994, **133**,1-3, 295-298.
280. Cheng Zhao-hua, Mao Ming-xi, Yang Chun-li, Zhang Yi-de, Li Fa-shen, Shen Bao-gen, Zhang Jun-Xian, Sun Ji-jun, J. Appl. Phys., 1994, **76**, 5, 2981-2985.
281. Barandiaran J.M., Gorria P., Orue I., Fdez-Gubieda M.L., Plazaola F., Hernando A., Phys. Rev. B, 1996, **54**, 5, 3026-3029.
282. Gu B.X., Xue D.S., Shen B.G., Li F.S., J. Magn. and Magn. Mater., 1997, **167**, 1-2, 105-108.
283. Munteanu C., Rusu I., Jianu A., Bul. Inst. Politehn. Iasi. Sec. 9. 1998, **44**, 1-4, 35-37.
284. Mushnikov N.V., Zajkov N.K., Serikov V.V., Kleinerman N.M., Gaviko V.S., Yermakov A.Ye., J. Magn. and Magn. Mater., 1997, **167**, 1-2, 93-98.
285. Schmool D.S., Garitaonandia J.S., Gorria P., Barandiaran J.M., J. of Magn.

and Magn. Mater., 1998, **177**, 1, 955-956.

286. Kopcewicz M., Grabias A., Idzikowski B., J. Magn. Mater., 1998, **177-181**, pt. 2, 1434-1436.
287. Slawska-Waniewska A., Greneche J.M., Inone A., J. Magn. Soc. Jap., 1999, **23**, 4, pt. 1, 194-196.
288. Carbucicchio M., Lanotte L., Palombarini G., Rateo M., Hyperfine Interactions, 2000, **126**, 1-4, 35-39.
289. Greneche J.M., Slawska-Waniewska A., J. Magn. and Magn. Mater., 2000, **215-216**, 264-267.
290. Negri D., Yavari A.R., Deriu A., Acta Mater., 1999, **47**, 18, 4545-4554.
291 Yang Y.Z., Zhu Y.L., Li Q.S., Ma X.M., Dong Y.D., Wang G.M., Wei S.Q., Physica B, 2001, **293**, 3-4, 249-259.
292. Ren Hong, Ryan D.H., J Appl. Phys., 1993, **73**, 10, pt. 2A, 5494-5496.
293. Gurevich G.M., Erzinkyan A.L., Parfenova V.P., Reiman S.I., Topalov S.V., Trchlik M., Pavlov V.N., Izv. AN, Ser. Fiz., 1994. **58**, 4, 97-100.
294. Vincze I., Kaptas D., Kemeny T., Kiss L.F., Balogh J., Phys. Rev. Lett., 1994, **73**, 3, 496-499.
295. Vincze I., Kaptas D., Kemeny T., Kiss L.F., Balogh J., J. Magn. and Magn. Mater., 1995, **140-144**, 1, 297-298.
296. Zamora Ligia E., Perez Alcazar G.A., Bohorquez A., Tabares J.A., J. Magn. and Magn. Mater., 1994, **137**, 3, 339-342.
297. Makhlouf S.A., Nakamura T., Shiga M., J. Magn. and Magn. Mater., 1994, **135**, 3, 257-264.
298. Al-Omari I.A., Jaswal S.S., Fernando A.S., Sellmyer D.J., Hamdeh H.H., Phys. Rev. B, 1994, **50**, 17, 12665-12671.
299. Godovikov S.K., Zolotukhin O.A., Ivanova T.I., Nikitin S.A., Rusakov V.S., Sergeev V.V., Tereshina I.S., Izv. AN. Ser. Fiz., 1994, **58**, 4, 146-150.
300. Sato T., Ando T., Oku T., Morimoto S., Ito A., J. Magn. and Magn. Mater., 1995, **140-144**, 3, 1781-1782.
301. Fiorani D., Lanotte L., Ronconi F., J. Magn. and Magn. Mater., 1992, **104-107**, 1, 141-142.
302. Pankhurst Q.A., Jiang J.Z., Betteridge S., Gibbs M.R.J., Gehring G.A., J. Phys.: Condens. Matter, 1995, **7**, 49, 9571-9593.
303. Zawora L.E., Perez Alcazar G.A., Tabares J.A., Bohorquez A., Marco J.F., Gonzaler J.M., J. Phys. Condens. Matter, 2000, **12**, 5, 611-621.
304. Schwer P.J., Celinsh Z., Heinrich B., J. Appl. Phys., 1994, **75**, 10, pt. 2A, 5583-5585.
305. Sakamoto I., Koguma K., Nawate M., Honda S., J. Magn. and Magn. Mater., 1997, **165**, 1-3, 208-211.
306. Mibu K., Passamani E.C., Elmassalami M., Shinjo T., Baggio-Saitovitch E., J. Magn. and Magn. Mater., 1995, **140-144**, 1, 623-624.
307. Richomme F., Teillet J., Auric P., Veillet P., Fnidiki A., Houdy Ph., Boher P., J. Magn. and Magn. Mater., 1995, **140-144**, 1, 627-628.
308. Fnidiki A., Juraszek J., Teillet J., Richomme F., Lebertois J.P., J. Magn. and Magn. Mater., 1997, **165**, 1-3, 405-407.
309. Ramamurthy Acharya B., Piramanayagam S.N., Nigam A.K., Shringi S.N., Prasad Sh., Venkataramani N., Chandra G., Krishnan R., J. Magn. and Magn. Mater., 1995, **140-144**, 1, 555-556.
310. Carbucicchio M., Palombarini G., Rateo M., Ruggiero G., Phil. Mag. B, 1997, **76**, 4, 425-432.
311. Kopcewicz M., Jagielski J., Stobiecki T., Gawlik G., J. Appl. Phys., 1994,

76, 9, 5232-5241.
312. Neubauer M., Lieb K.P., Schaaf P., Uhrmacher M., Thin Solid Films, 1996, **275**, 1-2, 69-72.
313. Shringi S.N., Piramanayagam S.N., Prasad Shiva, Venitramani N., Patni M.Y., Krishnan R., Tessier M., J. Appl. Phys., 1993, **73**, 10, pt. 2B, 6438-6440.
314. Sajieddine M., Bauer Ph., Cherifi K., Dufour C., Marchal G., Camley R.E., Phys. Rev. B, 1994, **49**, 13, 8815-8820.
315. Kuz'mak O.M., Shevchenko V.I., Pogorelyi A.N., Ukr. Fiz. Zh., 1994, **39**, 1, 81-85.
316. Kuzmak O., Pogorily A., Shevchenko V., J. Magn. and Magn. Mater., 1995, **148**, 1-2, 36-37.
317. Ruckert T., Tappert J., Brand R.A., Keune W., J. Magn. and Magn. Mater., 1997, **165**, 1-3, 411-413.
318. Sajieddine M., Bauer Ph., Marchal G., Nourreddine A., Ann. Chim. Sci. Mater., 1998, **23**, 3, 511-515.
319. Sajieddine Mohammed, Bauer Philippe, Marchal Gerard, Nourreddine Abdelmajid, C.R. Acad. Sci. Ser. 2, Fasc. B, 1999, **327**, 5, 523-527.
320. Hanisch K., Keune W., Brand R.A., Binek C., Kleemann W., J. Appl. Phys., 1994, **76**, 10, pt. 2, 6528-6530.
321. Roig A., Zhang X.X., Zuberek R., Tejada J., Molins E., J. Magn. and Magn. Mater., 1995, **140-144**, 1, 559-560.
322. Ono T., Hosoito N., Shinjo T., J. Phys. Soc. Jap., 1994, **63**, 8, 2874-2877.
323. Bauer Ph., Klose F., Schulte O., Felsch W., J. Magn. and Magn. Mater., 1994, **138**, 1-2, 163-172.
324. Bauer Ph., Andrieu S., Lemine O.M. Piecuch M., J. Magn. and Magn. Mater., 1997, **165**, 1-3, 220-223.
325. Emoto T., Mibu K., Hosoito N., Shinjo T., J. Phys. Soc. Jap., 1994, **63**, 9, 3226-3229.
326. Kawawake Y., Mibu K., Shinjo T., J. Phys. Soc. Jap., 1994, **63**, 7, 2700-2705.
327. Cui F.Z., Wang Y., Cui H., Li W.Z., Fan Y.D., J. Phys. D, 1994, **27**, 11, 2246-2250.
328. Li M., Ma X.D., Peng C.B., Zhao J.G., Mei L.M., Gu Y.S., Chai W.P., Mai Z.H., Shen B.G., Liu Y.H., Dai D.S., Phys Rev. B, 1994, **50**, 14, 10323-10326.
329. Li M., Ma X.D., Peng C.B., Zhao J.G., Mei L.M., Liu Y.H., Gu Y.S., Chai W.P., Mai Z.H., Shen B.G., Dai D.S., J. Appl. Phys., 1995, **77**, 8, 3965-3970.
330. Ustinov V.V., Tsurin V.A., Romashev L.N., Ovchinnikov V.V., Pis'ma v ZhTF, 1999, **25**, 11, 88-94.

Chapter 6

331. Bugaev S.P., Iskol'dskii A.M., Mesyats G.A., Proskurovskii D.I., ZhTF, 1967, **371**, 12, 2206.
332. Mesyats G.A. (ed.), High-current pulsed electron beams and technologies, Novosibirsk, Nauka, 1983.
333. Shchanin P.M. (ed.), Sources of charged particles with a plasma emitter, Ekaterinburg, Nauka, 1993.
334. Koval N.M., Oks E.M., Shchanin P.M., et al., Nucl. Instr. and Methods Phys. Res., 1992, **A321**, 417.
335. Thompson D.A., Radiation Effects and Defects in Solids, 1981, **56**, 105-150.

336. Russel H., Ruge I., Ion implantation, translated from English, edited by Guseva M.I., Moscow, Nauka, 1983.
337. Didenko A.N., Ligachev A.E., Kurakin I.B., Effect of beams of charged particles on the surface of metals and alloys, Moscow, Energoatomizdat, 1987.
338. Tsurin V.A., Sorkin A.M., Pavlov V.A., Stepanov A.P., Poverkhnost': Fizika, Khimiya, Mekhanika, 1988, 11, 131.
339. Didenko A.N., Rjabchikov A.I., Isaev G.P. et al., Mater. Sci. Eng., 1989, **A115**, 337.
340. Kreindel Yu.E., Ovchinnikov V.V., Vacuum, 1990, **42**, 1/2, 81.
341. Zhukov V.P., Demidov A.V., Uspenskii M.N., Phys. Stat. Sol., 1986, **B136**, 2, K91.
342. Zhukov V.P., Demidov A.V., Atom. Energiya, 1985, **59**, 1, 29.
343. Chudinov V.G., Cotterill R.M.G., Andreev V.V., Phys. Stat. Sol.(a), 1990, 122, 111.
344. Martynenko Yu.V., Yavlinskii Yu.N., Poverkhnost'. Fizika, Khimiya, Mekhanika, 1988, 6, 5.
345. Martynenko Yu.V., Yavlinskii Yu.N., ZhTF, 1990, **58**, 6, 1164.
346. Kamzin A.S., Grigor'ev L.A., Pis'ma v ZhTF, 1990, **16**, 16, 38.
347. Gol'danskii V.I. and Gerber G.M. (editors), Chemical application of Mössbauer spectroscopy, Moscow, Mir, 1970.
348. Litvinov V.S., Karakishev S.D., Ovchinnikov V.V., Nuclear gamma-resonance spectroscopy of alloys, Moscow, Metallurgiya, 1982.
349. Gel'd P.V., Levchenko V.P., Ovchinnikov V.V., et al., Determination of the interstitial positions of hydrogen atoms in metals by NGR method, In: Heat treatment and physics of metals, Sverdlovsk, 1987, 12-16.
350. Ovchinnikov V.V., Gel'd P.V., FMM, 1988, **65**, 1, 23.
351. Wertheim H., Hausmann A., Zander W., Electronic structure of point defects, translated from English, Moscow, Atomizdat, 1977.
352. Gonser U., Proc. Intern. Conf. on Mössbauer Spectroscopy, Cracow, 1975, **2**, 113 (cited after [356]).
353. Vogl G., J. Phys. (Paris) Collog., 1974, **35**, 6 (cited after [356]).
354. Dezsi I., J. Phys. (Paris) Collog., 1980, **41**, 1 (cited after [356]).
355. Sawicka B.D., Nuclear Physics Methods in Materials Research (eds. K.Bethge et al.), Vieweg, Braunschweig, Wiesbaden, 1980, 216 (cited after [356]).
356. Savitskaya B., Savitskii J., Study of implantation of iron ions by conversion electron Mössbauer spectroscopy. Mössbauer spectroscopy. Unusual applications of the method. Translated from English, edited by Gonzer U., Moscow, Mir, 1984.
357. Sprouse G.D., Kalvius G.M., Hanna S.S., Phys. Rev. Lett., 1968, **18**, 1041.
358. De Barros F.S., Hafemeister D., Viccaro P.J., J. Chem. Phys., 1970, **52**, 2865.
359. Zhukov V.P., Demidov A.V., Uspenskii M.N., Phys. Stat. Sol. (b), 1986, **136**, K91.
360 Yoshida Y., Nasu Y., Fujita F.E. et al., J. Magn. and Magn. Mater., 1983, **31/34**, 753.
361. Ali-zade I.I., Binnatov K.G., Gruzin P.L., et al., Izv. AN SSSR, Ser. Fizicheskaya, 1986, **50**, 12, 2304.
362. Sawicka B.D., et al., Hyperfine Interactions, 1978, **5**, 2,147.
363. Menningen M., Sielemann R., Vogl G., et al., Hyperfine Interactions, 1987, **35**, 1-4, 807.
364. Lewis S.J., Flinn P.A., Phys. Stat. Sol., 1968, **26**, 1, K51.

365. Schwalbach P., Laubach S., Hartick M., et al., Phys. Rev. Lett., 1990, **64**, 11, 1274.
366. Prasad K.G., Hyperfine Interactions, 1987, **34**, 1-4, 197.
367. Zhang Guilin, Liu Lianfan, Zhou Weinan, et al., Nucl. Tech., 1986, **12**, 1-5, 10.
368. Kolawa E., Tyliszcak T., Sawicka B.D., et al., Phys. Stat. Sol., 1987, **A102**, 1, 181.
369. Reuther H., Nucl. Instrum.and Meth. Phys. Res. B, 1991, **53**, 2, 167.
370. Bogdanov V.V., Bykov V.N., Zakurkin V.V., Klochko G.G., Metallofizika, 1988, **10**, 2, 112.
371. Bogdanov V.V., Zakurkin V.V., Petrov D.V., Pen'kov Yu.P., Study of proton-irradiated platinium by Mössbauer spectroscopy, In: Radiation material science, Proceedings of the International conference, Alushta, May 22-25, 1990, Khar'kov, **5**, 70, 221.
372. Marest G., Perez A., Gerard P., Mackowski J.M., Phys. Rev. B: Condens. Matter., 1986,. **34**, 7, 4831.
373. Bashkirov Sh.Sh., Ivoilov N.G., Romanov E.S., et al., Izv. AN SSSR. Ser. Fizicheskaya, 1990, **54**, 9, 1769.
374. Muramatsu H., Miura T., Nakahara H., Phys. Rev. B, 1990, **42**, 1A, 43.
375. Bakharev O.G., Pogrebnyak A.D., Martynenko V.A., Zekka A., Bruza R., Oshner O., Rissel H., Poverkhnost': Fizika, Khimiya, Mekhanika, 1995, 4, 52-66.
376. Toriyama T., Moser N., Kruijer S., et al., Nucl. Instrum. and Meth. Phys. Res. B, 1997, **129**, 4, 474-482.
377. Deweerd W., Moons R., Milants K., et al., Nucl. Instrum. and Meth. Phys. Res. B, 1997, **127-128**, 307-310.
378. Dezsi I., Szucs I., Fetzer Cs., Langouche G., et al., J. Phys.: Condens. Matter (J. Phys. F), 1997, **9**, 2, 579-584.
379. Kreindel' Yu.E., Ovchinnikov V.V., Fizika i Khimiya Obrabotki Materialov, 1991, 3, 14.
380. Ibragimov Sh.Sh., Kirsanov Yu.V., Pyatiletov Yu.S., Radiation damage in metals and alloys, Moscow, Energoatomizdat, 1985.
381. Bi Si-yun, Zhang Yue-Lu, Mei Lliang-mo, J. Phys., 1988, **49**, 12, Suppl. 2, 1365.
382. Reuther H., Rauschenbach B., Richter E., Vacuum, 1988, **38**, 11, 967.
383. Reuther H., Richter E., Fujinami M., Ujihira Y., Conversion electron Mössbauer spectroscopic studies of iron implanted with boron, carbon and phosphorus. Energy Pulse and Particle Beam Modif. Mater.: Intern. Conf. Sept. 7-11, 1987, Berlin, 1988, 231.
384. Uglov V.V., Fedotova Yu.A., Kuleshov A.K., Khodasevich V.V., Perspektivnie Materialy, 1998, 5, 67-72.
385. Uglov V.V., Fedotova Yu.A., Khodasevich V.V., Stanek Ya., Poverkhnost': Rentgen. Sinkhrotron. i Neitron. Issled, 1999, 5-6, 73-79.
386. Reuther H., Isotopenpraxis, 1988, **24**, 11, 12, 419.
387. Aimanov M.Sh., Zhetbaev A.K., Kadyrzhanov K.K., et al., Poverkhnost': Fizika, Khimiya, Mekhanika, 1988, 10, 134.
388. Marest G., El Khakani M.A., Nucl. Instrum. and Meth. Phys. Res. B, 1991, **59-60**, pt. 2, 833.
389. Zhantikin T.M., Zhetbaev A.K., Vereshchak M.F., Serikbaeva Z.T., Phys. Stat. Sol., 1987, **A101**, 2, 361.
390. Ramos S.M.M., Amaral L., Marest G., et al., Radiat. Eff. and Defects Solids,

1989, **110**, 34, 355.
391. Behar M., Amaral L., Ramos S.M.M., et al., J. Appl. Phys., 1990, **68**, 9, 4487.
392. Rusakov V.S., Kadyrzhanov K.K., Turkebaev T.E., Aimanov M.Sh., Zhukov V.N., Poverkhnost': Fiz. Khimiya. Mekh., 1995, 7-8, 28-35.
393. Rusakov V.S., Kadyrzhanov K.K., Turkebaev T.E., Aimanov M.Sh., Zhukov V.N., Poverkhnost': Fiz. Khimiya. Mekh., 1996, 11, 80-90.
394. Reuther H., Dobler M., Surface and Interface Anal., 1996, **24**, 6, 411-415.
395. Reuther H., Dobler M., Appl. Phys. Lett., 1996, **69**, 21, 3176-3178.
396. Kruijer S., Keune W., Dobler M., Reuther H., Appl. Phys. Lett., 1997, **70**, 20, 2696-2698.
397. Desimoni J., Sánchez F.H., Fernández van Raap M.B., Lin X.W., Bernas H., Cleerc C., Phys. Rev. B, 1996, **54**, 18, 12787-12792.
398. Dobler M., Reuther H., MQ´ller W., Hyperfine interactions, 1998, **112**, 185-188.
399. Noetzel J., Meyer D.C., Tselev A., Mücklich A., Paufler P., Prokert E., Wieser E, Möller W., Appl. Phys., 2000, **A71**, 47-74.
400. Reuther H., Betzl M., Watz W., Richter E., Hyperfine interactions, 1998, **112**, 165-168.
401. Reuther H., Betzl M., Watz W., Richter E., Hyperfine interactions, 1998, **113**, 391-401.
402. Reuther H., Hyperfine interactions, 1998, **111**, 135-140.
403. Kruier S., Nikolov O., Keune W., Reuther H., Weber S., Scherrer S., J. Appl. Phys., 1998, **84**, 12, 6570-6581.
404. Jagielski J., Marest G., Moncoffre N., Nucl. Instrum. and Meth. Phys. Res. B, 1997, **122**, 3, 575-578.
405. Johnson E., Johansen A., Sarholt-Kristensen L., et al., Nucl. Instrum. and Meth. Phys. Res., 1987, **B19-20**, 1, 171.
406. Sakamoto Isao, Hayashi Nobuyuki, Furubayashi Bunji, Tanoue Hisao, J. Metal Finish. Soc. Jap., 1988, **39**, 10, 652-655.
407. Sakamoto I., Hayashi N., Bull. Electrotechn. Lab., 1989, **53**, 11, 1-10.
408. Johnson E., Grabak L., Johansen A., et al., Nucl. Instrum. and Meth. Res. B, 1989, **39**, 1-4, 567.
409. Leutenecker R., Wagner G., Lousis T., et al., Mater. Sci. and Eng. A, 1989, **115**, 229.
410. Tsurin V.A., Sorkin A.M., Filippova N.P., Pavlov V.A., Effect of the dose of irradiation with low energy protons on the processes of redistribution of atoms in FeNiMn alloys, In: Modification of the properties of structured materials by charged particle beams, 2nd National conference proceedings, Sverdlovsk, 1991, 70.
411. Johnson E., Structural Phase Transitions in Ion Implanted Nickel and Stainless Steel. Thesis for the Degree of Dr. Sci. at the University of Copenhagen. Copenhagen, 1992.
412. Li J., Liu B.X., Phys. Stat. Sol., 1987, **A99**, 1, K31.
413. Hans M., Frech G., Wolf G.K., Wagner F.E., Nucl. Instrum. and Meth. Phys. Res.B, 1991, **53**, 2, 161.
414. Neubayer M., Lieb K.P., Schaaf P., Uhrmacher M., Thin Solid Films, 1996, **275**, 1-2, 69-72.
415. Crespo-Sosa A., Schaaf P., Bolse W., et al., Phys. Rev. B, 1996, **53**, 22, 14795-14805.
416. Amaral L., Scorzelli R.B., Paesano A., Brückman M.E., Bustamante Dominguez

Angel, Shinjo T., Ono T., Hosoito N., Surface Sci., 1997, **389**, 1-3, 103-108.

417. Bauer Ph., Dufour C.; Jaouen C., Marchal G., et al., J. Appl. Phys., 1997, **81**, 1, 116-125.
418. Richome F., Teillet J., Fnidiki A., Toulemonde M., Nucl. Instrum. and Meth. Phys. Res. B, 1977, **122**, 3, 507-509.
419. Santos D.L., De Souza J.P., Ameral L., Boudinov H., Nucl. Instrum. and Meth. Phys. Res. B, 1995, **103**, 1, 56-59.
420. Kopcewicz M., Jagielski J., Grabias A., et al., Nucl. Instrum. and Meth. Phys. Res. B, 1997, **127-128**, 141-144.
421. Borodin S.N., Kreindel' Yu.E., Mesyats G.A., Ovchinnikov V.V., Shabashov V.A., Pis'ma v ZhTF, 1989, **15**, 17, 51.
422. Kreindel' Yu.E., Mizgulin V.N., Ovchinnikov V.V., Phase transformations in Fe-Ni alloys under the effect of ion beams, In: Proceedings of the 2nd National conference: Modification of properties of structural materials by charged particle beams, Sverdlovsk, 1991.
423. Ovchinnikov V.V., Self-propagating phase transformations in metastable media induced by ion bombardment. Proc.XVIth Intern. Sympos. Discharges and Electrical Insulation in Vacuum. Moscow – St. Petersburg: 1994, (SPIE, v. 2259).
424. Ovchinnikov V.V., Phase transitions in metastable media induced by ion bombardment, In: Proceedings of 3rd International conference "Radiation physics of solids", Sevastopol, 1993.
425. Ovchinnikov V.V., Kogan Yu.D., Gavrilov N.V., Shtoltz A.K., Surf. and Coat. Technol., 1994, **64/1**, 1.
426. Borodin S.N., Kreindel' Yu.E., Mesyats G.A., Ovchinnikov V.V., Pis'ma v ZhTF, 1989, **15**, 13, 87.
427. Goloborodsky B.Yu., Ovchinnikov V.V., Semenkin V.A., Fusion Technology, May 2001, **39**, 1217-1228.
428. Ovchinnikov V.V., Chernoborodov V.I., Ignatenko Yu.G., Nucl. Instr. and Meth. Phys. Res. B, 1995, **103**, 313-317.
429. Syutkin N.N., Ovchinnikov V.V., Talantsev E.F., Chernoborodov V.I., Phil. Mag. A, 1994, **70**, 3, 439.
430. Ovchinnikov V.V., Russian Metallurgy, 1996, 6, 90-113.
431. Bischoff L., Teichert J., J. Phys. D: Appl. Phys., 2000, **33**, L69-L71.
432. Hausmann S., Bischoff L., Teichert J., Voelskow M., J. Appl. Phys., 2000, **87**, 1, 57-62.
433. Kamzin A.S., Grigor'ev L.A., Zhurnal Tekhnicheskoi Fiziki, 1990, **60**, 7, 151-156.
434. Andreeva M. A., Irkaev S.M., Prokhorov K.A., Salaschenko N.N., Semenov V.G., Chumakov A.I., Ruffer R., J. Alloys and Compounds, 1999, **12**, 322-332.
435. Nagy D.L., Baron A.Q.R., Bottyan L., Deak L., Dekoster J., Langouche G., Ruffer R., Semenov V.G., Sziilagyi E., Alloys and Compounds, 1999, **286**, 307-321.
436. Nagy D.L., Bottyan L., Deak L., Dekoster J., Langouche G., Semenov V.G., Spiering H., Sziilagyi E. Synchrotron Mössbauer Reflectometry in Material Science, High Technologies of the NATO Science Series, 1999, **13**, 128-139.
437. Semenov V.G., Nuclear gamma resonance optics of the surface and low-dimension laminated structures, In: Theses for the Dr. Sci. Degree, Institute of Analytical Instrument Making, RAS, St. Petersburg, 1999.

Appendices

438. Vasil'ev F.P., Lecturers on the methods of solving extreme problems, Moscow, Moscow State University, 1974.
439. Dem'yanov V.F., Rubinov A.M., Approximate methods of solving extreme problems, Leningrad, Leningrad State University, 1968.
440. Study of technological objects and their control using computers, Collection of algorithms in ALGOL-60, edited by V.V. Boiko, Moscow, NIITEKhIM, 1969.
441. Hooke R., Jeeves T.A. Journ. of Assoc Comp. Math., 1961, **8**, 2, 212-229 (cited after [208]).
442. Sveshnikov A.L., Fundamentals of error theory, Leningrad, Leningrad State University, 1972.
443. Ivanov V.K., Tanana V.P., Vasin V.V., Theory of linear ill-posed problems and its applications, Mocow, Nauka, 1978.
444. Tikhonov A.N., Arsenin V.Ya., Methods of solving ill-posed problems, Moscow, Nauka, 1979.
445. Vapnik V.S., Restoration of relationships using empirical data, Moscow, Nauka, 1979.
446. Window B., J. Phys. E: Sci. Insnrum., 1971, **4**, 5, @•. 401-402.
447. Hesse J., Rubartsch A., J. Phys. E: Sci. Instrum., 1974, **7**, 7, 526-532.
448. Ovchinnikov V.V., Sidorenko F.A., Pervukhin N.A., et al., FTT, 1981, **23**, 2, 617-621.
449. Tanana V.P., Methods of solving operator equations, Moscow, Nauka, 1981.
450. Mathematical encyclopaedic dictionary, Ed. By Prokhorov Yu.V., Moscow, Sov. Entsiklopediya, 1988.
451. Collection: Ill-posed problems of natural science, edited by Tikhonov A.N., Goncharskii A.V., Moscow, Moscow State University, 1987.
452. Groetsch C.W., Inverse Problems of the Mathematical Sciences, Wierweg, Braunschweg-Wiesbaden, 1993.
453. Vasin V.V., Ageev A.L., Ill-posed problems with apriori information, Ekaterinburg, Ural'skaya Izdatel'skaya Firma "Nauka", 1993.
454. Katsnel'son A.A., Alimov Sh.A., Dazhaev P.Sh., et al., FMM, 1968, **26**, 6, 987-995.
455. Katsnel'son A.A., Izv. VUZov. Fizika, 1974, **4**, 111-114.
456. Goloborodsky B.Yu., Ovchinnikov V.V., Semionkin V.A., Fusion Technology, 2001, **39**, 3, 1217-1228.

Index